HF Communications: A Systems Approach

Nicholas M. Maslin

MA, PhD, CEng, MIERE

Principal Consultant, Software Sciences Ltd.

Pitman

PITMAN PUBLISHING
128 Long Acre, London WC2E 9AN

© N M Maslin 1987

First published in Great Britain 1987

British Library Cataloguing in Publication Data

Maslin, N.M.
    HF communications : a systems approach.
    1. Telecommunication systems
    I. Title
    621.38   TK5101

    ISBN 0-273-02675-5

ISBN 0 273 02675 5

Printed at The Bath Press, Avon

Contents

Preface

Communications using the high frequency (HF) spectrum between 2 and 30 MHz have been experiencing a considerable resurgence of interest in the past few years despite having been virtually ignored in the 1960s and early 1970s in favour of satellite systems. Much of the expertise of skilled operators has been lost and with it an appreciation of many of the physical principles and characteristics upon which system judgements should be based. Now, with the advent of powerful microcomputers and very large scale integration (VLSI) technology, there is a tremendous potential to overcome many of the traditional problems that have plagued the HF communications system designer.

The aim of this book is to provide a firm foundation for the design, evaluation and operational use of HF communications. The subject is approached from the systems viewpoint; it is the systems issues relating to communications within the HF band that are primarily of concern.

It is intended that the book will be of use to a wide range of readers: to the professional communications engineer, already familiar with many of the problems but requiring a consolidated systems view of the subject: to the engineering student and research worker as a reference text to demonstrate how modern technology can be used to solve problems posed to the HF user: to non-specialist readers who wish to broaden their understanding of the physical principles, limitations and constraints upon an HF communications system.

The approach provides an understanding of the way in which various elements of the system contribute to the overall HF communications link performance. Mathematics is introduced only where it is needed to clarify or illustrate key issues. In a single book it would not be possible to deal with all relevant topics in great depth. Many subjects, for example error control coding, are discussed purely from the HF communications viewpoint rather than as a general treatise. It is intended that the reader who wishes to pursue a particular topic further will consult some of the references cited.

N.M.M. July 1986

1 HF Radio : Past and Present

1.1 Early Developments

1.1.1 Origins of Long-range Communications

A violent North Atlantic storm buffetted the old Newfoundland barracks huts. Inside a young scientist with a telephone headpiece clamped to his ears sat amongst a variety of strange looking instruments. Several men stood watching in tense, silent anticipation. For a time nothing happened. Then, suddenly, the scientist raised his hand, listened for a moment and then handed the headphones to his assistants for them also to hear the clicks of the Morse letter S. The date was 12 December 1901; the scientist, Guglielmo Marconi. A telegraphic signal originating from Poldhu on the Cornish coast had been received more than 3000 kilometres away across the Atlantic Ocean using 120 metres of wire as a receiving antenna, kept aloft by a high flying kite. Long-distance 'wireless' communications had been established.

Marconi's success was the source of considerable discussion in scientific circles. How had the radio waves found their way so far around the curved surface of the Atlantic? It was natural to suppose that the phenomenon was one of diffraction. Physicists and mathematicians directed their attention towards solving the problem of the diffraction of waves by a spherical surface. Calculations showed, however, that the diffraction effect was quite inadequate to explain the observed bending of the waves. From the work of Clerk Maxwell and Hertz it was known that radio waves travelled in a straight line and could be deflected from their path only if they encountered something having electrical properties different from those of the ether.

In 1902 Arthur Kennelly in America and Oliver Heaviside in England independently postulated the existence of a conducting layer in the upper region of the Earth's atmosphere. They suggested that this layer might be able to deflect the radio waves and allow them to follow the curvature of the Earth. The actual manner in which such a layer would affect the propagation of radio waves was not understood; it was a number of years before such a theory was provided. Meanwhile further experiments were being conducted.

During a voyage on the US liner *Philadelphia* in 1902 Marconi received messages over distances of 1000 km by day and 3000 km by night. He was thus the first to discover the fact (without appreciating its physical significance) that transmission conditions at night can be rather different from those encountered during the day.

The first significant application for radio waves was ship-to-shore communications which employed frequencies in the medium waveband (about 500 kHz or 600 m in wavelength). Their value was demonstrated dramatically in 1909 and 1912 when the sinking passenger ships *Republic* and *Titanic* used radio to summon assistance. By 1910 radio messages between shore stations and ships had become commonplace and in that year the first air-to-ground radio contact was established. Users of the new communications medium multiplied rapidly; by 1918 over 5000 ships had wireless telegraphy installations.

Marconi continued to improve his basic devices, achieving greater and greater ranges by increasing the wavelength of operation. In 1910 he was able to receive long wave signals of around 3 km in wavelength at Buenos Aires from Clifton, Ireland, a distance of 9600 km. In 1918 a radio telegraph message from the long wave station at Caernarvon, Wales was received at a distance of 17700 km, in Australia.

The remarkable progress made with long waves using lower and lower frequencies to achieve greater and greater ranges of practical communication distracted nearly all attention and research effort from the short waveband. Medium wave transmissions were known to have a comparatively short effective range during the day; even at night when their range was increased significantly their signal strength decreased at a much greater rate than that of long wave carriers. As a result it was assumed that the longer the wavelength the better the propagation.

1.1.2   Experiments with the Short Waveband

The assignment of frequencies above 1500 kHz (less than 200 m wavelength) to amateurs in the early years of radio was made in the firm belief that these frequencies were of little value for long-distance communication purposes. A short wavelength transmission of 2.5 MHz (120 m) had been employed for communications between ships as early as 1901. It had been noted that reception over distances of 1600 km or more could be obtained, but this was thought to be caused by freak conditions.

The conducting upper atmosphere, though postulated by Kennelly and Heaviside in 1902, lacked direct evidence for its existence during the following twenty years. Agreement of theoretical calculations with quantitative measurements on long wave field strength and the many peculiar features of short wave propagation discovered in Marconi's experiments since 1921, indicated the existence of such a layer. However, these phenomena were evidence of an indirect nature only.

The first direct evidence for the conducting Kennelly-Heaviside layer came in 1925 when Appleton and Barnett, by comparing the intensities of fading signals received simultaneously on a loop and on a vertical antenna, proved the presence of indirect or *sky waves*. The existence of more than one layer was indicated. Since these layers were believed to be composed of electrically charged particles, known as ions, the region became known as the *ionosphere*.

Experiments not only gave direct evidence of the reflecting layers within the ionosphere but also disclosed some of their properties, for example their heights. In 1926 Breit and Tuve showed that a wave packet or a pulse of small duration sent out from a transmitter produced two (sometimes more) impulses instead of one in a receiver placed a few kilometres from the transmitter. The

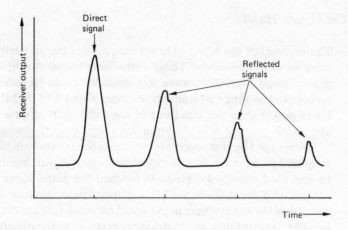

Fig. 1.1   *Receiver response to an HF signal*

obvious conclusion was that the impulse that arrived first was caused by the direct wave travelling along the ground; those that arrived later were a result of indirect waves or echoes reflected from the ionosphere (see Figure 1.1).

Ten years before these experiments Marconi began to experiment with short wave transmissions in the band 2–30 MHz, now known as the High Frequency (HF) band. This was not in an effort to achieve greater ranges since, as already mentioned, it was believed that the longer the wavelength the better the propagation; rather it was because the shorter wavelengths permitted directional antennas to be used to concentrate the transmitted energy into a narrow beam.

By 1922 Marconi was experimenting with high frequencies between Zandvoort in Holland and Southwold, Hendon and Birmingham in England. Remarkable differences in the behaviour of signals in the 5–10 MHz band over the different paths were noted. At Southwold, about 175 km from Zandvoort, the signal strength at night was much greater than that by day; at Birmingham, 450 km from Zandvoort, the signal strength by day appeared normally to be greater than that at Southwold while at night it was impossible to receive any sign of the signals. There seems little doubt that the day signal at Birmingham was caused by sky wave propagation, although Marconi failed to recognise it as such. It was to take him two more years to 'rediscover' it.

Further experiments were performed over much longer links with the aid of the steam yacht *Elettra*. These tests were performed initially with a frequency of about 3 MHz, using a 1 kW transmitter, between Cornwall and the Caribbean island of St Vincent, a distance of some 3700 km. It was found that these short wavelengths, of around 100 m, could be used to provide a satisfactory service at great distances. However this appeared to hold true only when the great circle path followed by the radio wave was all, or substantially all, in darkness.

1.1.3   HF Gains the Upper Hand

The reversal of the trend to lower frequencies began partly as a result of the work of radio amateurs. These enthusiasts found, to the amazement of the professionals, that short wave transmissions could be received over long distances. In the range of frequencies from about 5 to 30 MHz signals of much lower power than the hundreds of kilowatts used at long wavelengths could apparently provide reliable communications over thousands of kilometres.

Encouraged by this success in reversing the technological trend towards low frequencies, Marconi began experimenting with still higher frequencies than he had used previously. He soon reached the point where he could maintain contact with Australia on 9.8 MHz. Further successes followed. He was able to communicate to anywhere in the world he wished, at almost any time of the day or night, by careful choice of the appropriate operating frequency. It was at this time, during 1924, that the Marconi beam system was conceived.

The foundations were now laid for the Imperial wireless chain, a revolution in worldwide communication. This was a controversial plan, first raised in 1906, for linking the British Empire by a network of wireless communication stations. It had re-emerged after delays caused by the First World War and much political in-fighting.

By 1924, however, tests had proved the usefulness of high frequencies for long-distance communications. Marconi had demonstrated that the existing high-power low-frequency communications networks which were originally specified for the Imperial system were, by comparison, uneconomic and virtually obsolete for point-to-point communications. It was therefore agreed by the governments of the UK, Canada, Australia, South Africa and India to adopt the beam system on the basis of these tests. The age of HF was born.

By this time radio had already attracted numerous amateur enthusiasts and experimenters. Radio amateurs were achieving considerable success. For example, they maintained contact across the Atlantic during daylight throughout most of February 1925 with the aid of the shorter wavelength transmissions.

Running parallel to short wave communications experiments were activities in short wave broadcasting. The invention of the thermionic valve by the British scientist Sir Ambrose Fleming some years earlier had made wireless telephony possible. This resulted in a great interest in the possibility of broadcasting. The appetite for the new entertainment medium was insatiable and in 1927 the BBC decided to initiate short wave broadcasting to the Empire.

During the next few years radio telegraphy continued to be a primary means for ship-shore communications whilst radiotelephony was extended to a great variety of special applications. In 1927 a short wave telephone link from the British liner *Carinthia* sailing in the Pacific Ocean established contact with Britain. All passenger ships were soon equipped with radio for passenger use. The radio telephone was adopted by fishing fleets during the 1930s because it had the advantage of not requiring a skilled operator with knowledge of Morse code. After 1934 many small coastal craft were equipped for radio contact with shore.

Prior to the Second World War communications with aircraft were almost entirely by HF, and commonly employed hand-operated Morse, although the first trials of air-ground voice communications had been made in the US as early as 1928. The Second World War saw the development and introduction of VHF and UHF for airborne communications. However, since these frequency bands were constrained by line-of-sight propagation, any communications link to the aircraft beyond-line-of-sight continued to require the use of frequencies in the HF band.

Just after the end of the War, before VHF air traffic control communications were fully implemented, air-to-ground communications were conducted using frequencies between 3 and 6 MHz and ground-to-air communications at about 300 kHz. Where air-to-ground communications ranges of about 50 km were desired, communications were often found to be very poor, whereas the ground station could receive distant stations (greater than 1000 km) loud and clear. These problems, typical of airborne HF communications, are examined later in Chapter 7.

The HF band was now firmly established as the primary means of beyond line-of-sight communications. Pioneers of radio concentrated their long-range experiments in the HF band. This effort brought about a continuing stream of refinements in technology that, in turn, improved the level of attainable HF performance. Some of the more notable advances were:

improved vacuum tubes and antennas;
single sideband for more spectrum efficiency;
automatically tuned transmitters;
propagation prediction methods;
high-speed multitone data modems;
frequency synthesisers to replace crystals;
the introduction of solid-state circuits.

All these individual contributions in some way improved the link performance of the highly variable HF channel. In particular, the auto-tune and synthesised radios made it possible to change the frequency easily to follow the varying propagation conditions.

1.2   Changes in Fortune

1.2.1   The Reliance upon the Operator

Even during the 1960s HF equipment continued to need highly skilled communications specialists and operators who had considerable understanding of the transmission medium. It was only through sheer skill and years of experience that an operator could avoid poor-quality channels and identify the optimum frequency for a given path and time. Changing frequency was a long and tedious process, usually manual, and involved the use of different antennas which had to be matched to the transmitter.

With older-generation equipment the control of operating frequency or type of message traffic was carried out manually at the operator's discretion via a radio-set control unit. All outgoing calls and responses to incoming calls were initiated by the station operator, who had to monitor all the communications on the selected channel in order to respond to calls directed to the station. It was also the operator's responsibility not to interfere with traffic already in progress and to monitor other appropriate channels. This necessitated frequency and time co-ordination with other stations in the network.

All this required a high degree of concentration in order to maintain satisfactory communications. As a result, the level of operator frustration and fatigue was very often quite high. Despite these shortcomings, the HF operators of the day could nearly always provide highly reliable and good-quality links.

1.2.2   The Fall from Favour

In the early 1970s satellite communications emerged as an alternative to HF. Communicators weary of the endless pursuit of the ideal HF channel in the face of diurnal, seasonal and sunspot changes, eagerly boarded the satellite bandwagon. Suddenly HF seemed obsolete; it could be replaced by a reliable, nonvarying medium. Many military users switched to satellite as their primary communications medium, usually preserving the old HF equipment as a back-up.

Coincidentally, the early 1970s were leading into a sunspot minimum after a dismal sunspot maximum in 1968. Low sunspot numbers often mean low-quality and low-reliability HF communications. The frustration of communicators attempting to use the limited available HF propagation in the early 1970s may well have made the new satellite alternative seem that much more attractive.

Communications satellites were considered to be far superior to HF links, offering permanent clear and reliable worldwide communications. They had more voice channel capacity and could handle very high data rates. There was, however, one basic weakness in this type of link, namely the satellite itself. Whilst its invulnerability was never seriously in question during most of the 1970s, the satellite did present certain handicaps. Once deployed, it was inaccessible for maintenance and repair. Only a few countries, in particular the US and USSR, had the technical capability and the money to build satellites and place them in orbit. Consequently, most countries wishing to take advantage of this new mode of communications had to use satellites which were controlled by another country. When compared to the problems encountered in the HF band, particularly with older-generation equipment, the above weaknesses in satellite communications were found to be generally acceptable to most military and civilian users.

The advent of the communications satellite had two important effects upon the development of HF. Firstly, research and development of HF systems all but ceased during this period. Any technical progress that was made was relatively slow; many users believed that HF would become obsolescent. Secondly, the skills and experience traditionally associated with operators of the HF band were lost.

1.2.3   A Resurgence of Interest

Within a few years of the commencement of the satellite era, however, it was clear that satellites were not the panacea they at first appeared to be. Satellite communications for many civil applications, for example to replace traditional air traffic control, proved prohibitively expensive in the late 1970s. In particular the cost of the airborne installation has been a major factor in the abandoning of the AEROSAT programme. For many applications, therefore, HF will continue to remain the primary means of beyond-line-of-sight communications, mainly as a result of its lower cost and free provision by nature in the upper atmosphere of a reflecting region, albeit an imperfect one.

By the beginning of the 1980s progress in anti-satellite activity had been such as to make many experts believe that enemy action could put satellite links out of service very soon after the outbreak of hostilities. This could be achieved in a number of ways, for example by the destruction of the satellite itself with a missile, by electronic jamming or by damage to the satellite or its antennas by electromagnetic pulse (EMP) interference with the electronic circuits. Realization of the fact that satellite links, once considered to be inviolable, could be vulnerable has now led the military, particularly in the US, to reappraise its means of long-range communications.

At the same time that the satellite vulnerability was acknowledged, military strategic and tactical studies pointed to the communications scenario in force today: that no single, long-range communications medium is suitable for all requirements. Rather, successful command, control and communications (C^3) require a well-conceived mix of alternative communications links, one of which is HF.

These ideas, in particular the possible weaknesses in satellite links, have given a new lease of life to high frequency communications. Attention has once again turned to the HF spectrum.

1.3 Present Usage

1.3.1 HF Spectrum Allocation

Present-day uses of the HF spectrum, allocated to the 2–30 MHz band, cover a wide range of commercial and military applications. The occupancy of the spectrum and the users' demand for channel allocation is considerable.

Because radio space is potentially available to any user, international and national regulations have evolved to limit usage, available frequency bands and geographical limits of transmission. As a worldwide communications medium, HF is subject to the international radio regulations which have been developed by the International Telecommunications Union (ITU) in Geneva. These regulations include radio spectrum utilization and allocations within the HF band. Allocations are made on the basis of service type or usage as follows:

a) *Fixed* Radio communication between specified fixed points, i.e. point-to-point HF circuits.

b) *Mobile* Radio communication between stations intended to be used while in motion or during halts at unspecified points or between such stations and fixed stations.

c) *Aeronautical Mobile* Radio communication between a land station and an aircraft or between aircraft.

d) *Maritime Mobile* Radio communication between a coast station and a ship or between ships.

e) *Land Mobile* Radio communication between a base station and land mobile station or between land mobile stations.

f) *Broadcasting*   Radio communication intended for direct reception by the general public.

g) *Amateur*   Radio communication carried on by persons interested in the radio technique solely with a personal aim and without pecuniary interest.

h) *Standard Frequency*   Radio transmission of specified frequencies of stated high precision intended for general reception for scientific, technical and other purposes.

The allocation of these service types to the HF band is shown in Figure 1.2.

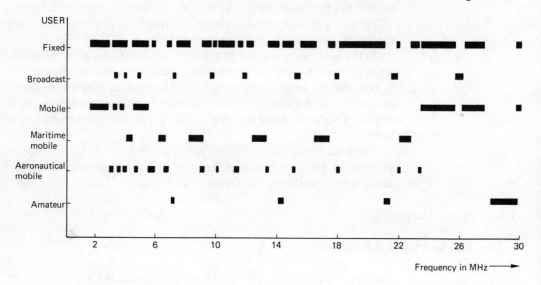

Fig. 1.2   *Frequency allocations in the HF band*

Fixed point-to-point communications, commonly used for international radio telephony and radio telegraphy is allotted a large proportion of the HF spectrum, as shown in Figure 1.2. Special procedures are required for shifting frequencies within the band because frequencies may become unusable as a result of changing propagation conditions.

HF radio telegraphy systems commonly use continuous wave (CW), frequency shift keying (FSK) or single sideband (SSB) transmission with subcarriers. The data transmitted includes Morse, teletype and other signals. SSB systems are employed for HF radio telephony. These systems use from 3 to 12 kHz bandwidth accommodating 1 to 4 voice channels respectively. In the 4 voice channel system, 2 channels are transmitted above and 2 below the carrier frequency. These voice channels can be used alternatively to accommodate a group of frequency division multiplexed telegraphy channels. In one technique, 3 SSB telegraph subcarriers (channels) each carrying 200 baud and requiring 340 Hz bandwidth are spaced 340 Hz apart to cover a portion of the 2465 Hz wide voice band. Each telegraph channel is in turn capable of accommodating four 50 baud teleprinter channels in time multiplexed form.

The high frequency portion of the spectrum has long been recognised as a useful and economic medium for achieving wide distribution of information over long distances. Although satellite communications systems are becoming more widely available and more attractive economically, HF will continue to be used extensively by many nations for a wide range of communications purposes.

1.3.2 Non-military Users

HF is used extensively to meet civilian and diplomatic communications requirements.

Telephone operations and private companies are major users of the fixed service. HF provides a means of communication to isolated or remote areas where alternative services are not available. The traffic that is carried is usually low in capacity; only limited resources are often provided by the user to improve the system. Many smaller companies use HF to avoid the cost of long-distance telephone charges of the carrier networks. The initial stages of development of remote mining or exploration operations often use HF radio in the period before a regular telephone service, using microwave relays or land lines, can be established.

Unless a satellite system is used, long-distance broadcasting is possible only by means of the HF band. Groups of channels are allocated throughout the HF band (see Figure 1.2) for the purposes of external broadcasting beyond the national boundaries of the originating authority. The number of channels available is now insufficient for the countries wanting to broadcast. In spite of a degree of international control interference is often a serious problem; it is particularly acute on the lower frequencies in the band at night when more channels are in use because of the prevailing ionospheric conditions. Frequencies in the region of 4 MHz are used in the tropics for local broadcasting. These have some advantage over medium waves during the tropical rainy season when atmospheric noise caused by thunderstorms is serious. However, the signal must be received by almost vertical reflection from the ionosphere; fading and distortion are thus generally greater than with medium waves.

An HF service is provided between embassies and missions in various countries. The radio facility is usually located in the embassy building itself with the antennas mounted on the roof. The United Kingdom Foreign and Commonwealth Office operates a communications network between the Foreign Office in London and every British embassy in the world. Although many of these links are by telephone, telex or public networks over satellite or cable, some 60 embassies are linked by an HF radio telegraphy network.

Commercial aircraft and shipping use HF extensively. Aeronautical mobile services are required by airlines that provide regional, national and international flights. For some of the smaller airlines providing mainly regional services within a country the radio equipment is inexpensive and performance tends to be marginal. Aeronautical land-based stations for airlines with national and international services are provided with much higher quality equipment; the HF service is not generally used, however, when other more reliable radio services are available within the area of operations. The HF mobile bands also provide communications between coast stations and ships and between ships. The ground wave mode of propagation enables communications to be established over long distances in coastal regions, whilst the sky wave provides communications from ships back to their shore base at distances up to halfway round the world.

The portions of the HF band reserved for amateur use are always congested. A recent resurgence of interest in amateur operations has been provided by the considerable growth in citizens' band radio, known as CB.

1.3.3   Military Users

Military forces use HF extensively for both strategic and tactical communications, either as a primary communications link or as a back-up. The bulk of the traffic is transmitted via sky wave although ground wave transmission is used extensively by some armies. Heavy reliance is still placed upon HF for use in land, sea and air operations.

The requirements of a military communications system vary from repetitive broadcasts of simple commands to secure digital communications. Global operations demand that the HF system adapt to propagation conditions in regions varying from equatorial to polar. Strategic HF links are used between national command headquarters (HQ) and forces dispersed around the world, as well as between Corps HQs and national or Allied HQs. Strategic HF links are also used between airborne command posts, early warning and reconnaissance aircraft and national command posts.

Other typical uses of HF radio systems on the battlefield include more tactical applications, for example:

a) Fixed point-to-point circuits, carrying high-priority voice and teleprinter traffic in case of failure of a primary trunk circuit such as tropospheric scatter.

b) Fixed point-to-point circuits where no other primary trunk circuit exists because of low traffic density.

c) Fixed to transportable circuits, between a fixed site and transportable shelters.

d) Fixed to mobile circuits.

e) Mobile to mobile circuits, particularly those between mobile elements operating beyond the reach of normal VHF nets.

Satellite communications are still too expensive to contemplate for general widespread use on the battlefield; their vulnerability is also a cause for concern. The traditional VHF/UHF links are unable to cope with beyond-line-of-sight communications requirements. For all practical purposes UHF is limited to line-of-sight and can be successfully operated only when there is visual contact. VHF, though better than UHF in this respect, also suffers from screening difficulties. The only reliable means of communications must make use of surface waves or short-range sky waves. Such communications can best be accomplished by use of the low frequency end of the HF band. For example man-portable HF-SSB transceivers are sometimes more useful than VHF sets, particularly in thick jungle or rough terrain. HF was extensively used in a high-angle sky-wave mode during the Vietnam conflict.

Although satellite terminals are becoming more common with some navies for long-range maritime communications, a considerable amount of the traffic is still passed via HF broadcast and ship-to-shore networks. Communications within a task force are mainly effected by HF ground wave; coordination with aircraft at long ranges is still an important role for HF sky waves.

In civil aircraft, where the use of the radio operator aboard airliners has been eliminated, speech is the normal mode of communication. Military aircraft, however, operate other modes besides speech. Morse is still a fallback position where a crew member has this skill; radio teletypes operating at low data rates are also in use in some air force applications where security is a requirement.

Air forces operate HF air-ground communications systems in a wide variety of operational roles; long-range maritime patrol aircraft operate extensively in northern waters above the Arctic Circle, encountering particularly difficult auroral propagation conditions; transport and tanker aircraft also operate at considerable ranges from base. Low-flying strike and reconnaissance aircraft operating over central Europe have much shorter communications range requirements to their bases but are nonetheless beyond-line-of-sight. Support helicopters pose still shorter communications requirements of tens of kilometres to the supported ground troops. The rugged nature of the terrain again necessitates the use of the HF band to effect contact with the tactical ground communications nets in these regions.

1.3.4   Evolving Technology

The wide range of users serves to illustrate the very diverse areas of application for HF communications; in each scenario a different set of problems is posed for the user and an equally diverse set of solutions may be appropriate.

Many of the HF radios still remaining in use were designed in the 1955–70 period and can now be considered obsolete. Some of these older units do not even cover the full 2–30 MHz HF spectrum, their operation being limited to the lower parts of the HF band. Many were designed to operate only in the single sideband (SSB) voice, amplitude modulation (AM), continous wave (CW) or 75 bits per second frequency shift keying (FSK) modes. They still use crystal oscillators and do not offer continuous tuning, whereas modern sets use highly accurate synthesised tuning increments of 100, 10 or even 1 Hz. There is thus much scope for technology innovation.

Progress made in HF technology and techniques since the late 1970s has been considerable. Today it is very difficult to find operators with HF skill and experience. The majority of future users will not be trained HF specialists. Military users will be operators who will have to perform other demanding tasks as well. Consequently, modern HF equipment incorporates a high level of automation and microprocessor control to relieve the operator workload and the reliance upon traditional human operator skills. There have also been developments in automatic tuning and antenna matching systems, remotely controllable systems, solid-state circuits and synthesised frequency selection systems. In addition, highly stable oscillators, frequency-agile synthesisers, fast tuning antenna couplers and solid-state power amplifiers are now available and offer the possibility of improving future HF communications significantly. The impact of very large scale integration (VLSI) will be evident in all types of future communications systems. It provides the ability to reduce size, weight and power consumption of existing equipment and to increase the overall system reliability. It also leads to the introduction of more complex systems by introducing other techniques and applications which have not previously been possible due to the lack of necessary technology.

The marriage of computer technology with HF communications systems offers tremendous promise to improve performance and reliability. The quantum leap in microprocessor technology has coincided with the resurgence of interest in the HF band, promising to revolutionise the HF world. The focus has been on using the best new technologies to overcome the traditional weaknesses of HF communications. In particular, the choice of a good channel has been automated by various systems, and further augmented by the use of robust data modulation and coding schemes which can tolerate a lower-quality channel. At the same time hardware evolution has reduced the size and cost of radio equipment, whilst greatly increasing its capability and performance. Some of the greatest advances are yet to come, however. The next decade holds the promise of breakthroughs in several areas of technology. These are described later in Chapter 10.

In order to understand fully the significance of future trends and developments it is necessary to study in more detail the fundamental principles of HF communications. This investigation must begin at the highest level, that is, in terms of a system description, the subject of Chapter 2.

2 System Considerations

2.1 Concepts

2.1.1 The Systems Approach

The concept of a *system* arises from the fact that assemblies of parts organised together in special ways can reveal unique properties which are not possessed by those same parts in isolation. A system possesses some particular property only when it exists as a coherent, organised entity; the property ceases to exist if the system is broken up into its component parts or if these parts are reorganised in a different manner. A system can therefore be understood fully only by studying the whole entity and not merely by analytical study of its constituent parts.

An investigation of systems requires *synthetic* rather than *analytic* thinking. In the analytic mode an understanding of a complete system is derived from an explanation of its parts; by contrast in the synthetic mode the system is viewed as part of a larger system and is explained in terms of its role within that larger system. The synthetic mode of thought, when applied to systems problems, is called the *systems approach*.

The systems approach is a relatively new phenomenon. The essential ideas underlying systems thinking are deceptively simple. However, the ideas have emerged from a variety of sources over a period of time; they have come together only gradually, in recent years.

2.1.2 System Characteristics

Systems may be of the closed or open type, as shown in Figure 2.1. It is important to draw a distinction between them. A *closed system* is a theoretical

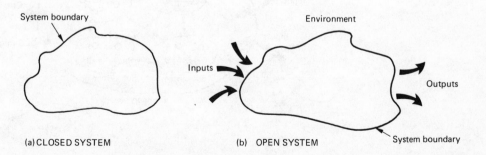

Fig. 2.1 *Closed systems and open systems*

concept which is impossible to achieve in the real world. It has rigid, impenetrable boundaries; there are no connections with other outside systems or influences.

Open systems, as implied by their name, are open to outside influences. Those influences which affect, but are not noticeably affected by, the system form the *environment* of the system. As Figure 2.1 shows, there is a line or envelope surrounding the system and separating it from this environment. This envelope is called the *system boundary*. An open system has permeable boundaries between itself and its environment.

In an open system there exist relationships between *inputs* from the system's environment and the system's own *internal properties*. There are also relationships between these internal properties and the response of the environment to the system's *outputs*. It is often convenient to treat the system as a 'black box', particularly when it cannot be analysed in a simple manner. In this approach (see Figure 2.2) it is unnecessary to understand the internal mechanisms of the system; inputs are merely transformed into outputs. By altering the level of the inputs whilst observing the effects on the outputs, sufficient information about the system can often be obtained.

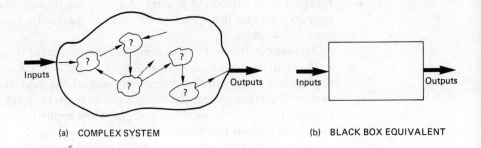

(a)   COMPLEX SYSTEM   (b)   BLACK BOX EQUIVALENT

Fig. 2.2   *Complex system viewed as a black box*

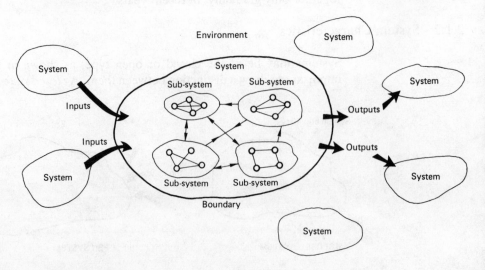

Fig. 2.3   *A system with sub-systems which interacts with other systems*

A system viewed *functionally* is an indivisible 'whole' in the sense that some of its essential properties are lost in taking it apart. Viewed *structurally* it is divisible into a number of parts or *sub-systems*. The sub-systems may themselves be systems in their own right just as every system may be part of a larger system, see Figure 2.3. However, a system cannot be divided into totally independent sub-systems since, by definition, it always has some characteristic or can display some behaviour that none of its sub-systems can.

System performance depends critically upon how the sub-systems fit and work together and not upon how each sub-system performs in isolation. System performance therefore depends upon interactions rather than actions; it depends upon how the system relates to its environment and to other systems within that environment. The systems approach necessitates an evaluation of the functioning of the system within the larger system in which it resides.

2.1.3   Design Considerations

Systems often have a number of different objectives so that some form of compromise is essential. Therefore a balance, or trade-off, must be sought between the conflicting objectives if the best overall result is to be obtained.

The fundamental principle of system design is to maximise the effectiveness of the system. This exercise must be done continually in the design process; designing to an unnecessarily tight specification may result in excessive cost, whilst cutting corners may compromise the desired performance. In a simple two-parameter trade-off analysis it may be quite simple to find a maximum. In a complex system there may be many tens or even hundreds of parameters to optimise. This leads to cost effectiveness studies in which the system performance is to be maximised for a fixed cost or, alternatively, the cost minimised for a fixed effectiveness. *System engineering* comprises the set of activities which leads to the creation of a complex man-made entity together with information flows associated with its operation. Because it is virtually impossible to find a single number which represents realistically the effectiveness of a complex system there is a good deal of art, as well as science, in system engineering.

The design of a system must begin with the users' requirements; these must be analysed carefully. Essential and desirable attributes of the system must be formulated and mapped on to the prevailing technologies and cost constraints. A top level design is then evolved that will result in the development of a service which meets the users' needs in full as far as is possible within the prevailing contraints.

It is particularly important that the fundamental objectives of the system should not be affected adversely in an endeavour to accommodate events of extremely low probability. System trade-offs must be assessed carefully; it is often the case that the last few percentage points of system performance are achieved by a disproportionate amount of the total system cost. Under these circumstances the systems analyst must question the requirement and seek further guidance from the user as to the motivation behind the stipulated system performance criteria. In many systems a compromise may be possible; for example rather than attempt to accommodate all possible outcomes automatically, however unlikely, the system could be designed to handle most events automatically and to sound an alarm calling for manual intervention when an uncommon event occurs which is beyond its capabilities.

The process of system design is composed of logical steps, but these steps are not always performed in order. Logically the problem must be formulated before it can be solved. However, the problem cannot be formulated adequately until it is well understood, and because it cannot be well understood until it has been more or less solved, the two are inseparable. Both *formulation* and *solution* therefore need to be performed simultaneously throughout the system design process.

The performance of a given sub-system interacts with the performance of other sub-systems. Sub-system design should not therefore be undertaken in isolation from those other sub-systems. The division of a system into sub-systems is somewhat arbitrary and is largely for convenience in administering the sub-system work. Generally a sub-system should be in a single geographic location and should totally include or totally exclude those interfaces or interconnections that cause most difficulty. If the division is done well, a team can be given responsibility for the design of a sub-system and work on it independently with only occasional feedback to other parts of the system design effort.

The optimisation of each sub-system independently will not, in general, lead to an optimum system. Furthermore, the improvement of a particular sub-system may actually worsen the overall system. Since every system is merely a sub-system of some larger system, this presents a difficult problem for the system designer. The heart of the matter lies in the complexity of the system and the danger of being unable to see the forest for the trees. The designer must cope with the various sub-systems and component parts in such a way as to optimise the cost effectiveness of the overall system.

The urgency of the need for an operational system often stimulates 'telescoping' of the time schedule for design and development; for example, prototype construction is undertaken before the design effort is completed, or production is started before the prototype has been tested. Such practices may lead to wasted time and effort, involving reworking of certain phases. However, this must be balanced against the equally obvious disadvantage of proceeding in an orderly fashion since a long development programme may imply that the system, when finally complete, may actually be too late to be useful. The system designer must cope with two conflicting trends: the rate of technological change speeds obsolescence; increasing system complexity stretches necessary development times.

The discussion so far has considered systems in very general terms. It is now necessary to apply the concepts that have been outlined to radio communications system design and to consider the specific problems and issues in more detail.

2.2   Radio Communications Systems

2.2.1   System Definition

Using the concepts discussed in the previous section it is possible to define a radio communications system in black box terms as shown in Figure 2.4. A message is input to the system from some source within the environment. This is operated upon by the processes within the system and subject to the environment in which it resides. A message is delivered to some output in a form which may or may not be meaningful according to the processes which it has undergone.

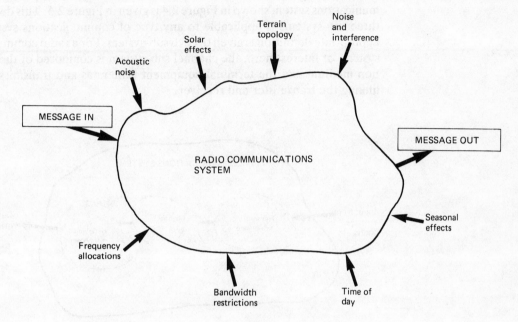

Fig. 2.4 *Radio communications system as a black box*

A typical radio communications circuit is required to perform a function. Usually this function involves the transmission of information from one location to another. The measurement of how well the circuit performs involves both the volume of information that can be transmitted during a given time interval as well as the accuracy with which the input information is reproduced at the output. The information can be of various forms. It is desirable to maximise information transfer rate and minimise intended radiation.

Nature and man are the sources of all engineering problems. Effects of nature are centred upon the existence of the Earth and its environment within the universe. Electromagnetic noise impinging upon the Earth from solar and cosmic sources establishes a bound on the information that can be conveyed throughout the radio spectrum. This bound is modified by radiation from noise sources within the troposphere, the terrestrial environment and, primarily, from man-made radio frequency sources and thermal noise. Long-term characteristics of this noise affect the required transmission power whilst short-term characteristics determine how the signal should be designed and detected to convey the desired information.

The above discussion relates to the system from the black box viewpoint. It is now necessary to consider how the system is structured in terms of its constituent sub-systems.

2.2.2   System Decomposition

It has already been stated that decomposing a system into sub-systems is essentially an arbitrary choice. A careful selection can simplify subsequent understanding. A convenient means of structurally decomposing the communications system shown in Figure 2.4 is given in Figure 2.5. This division into three sub-systems is applicable to any type of communications system if the appropriate definition is given of each sub-system. For a radio communications system, of interest here, the channel sub-system is composed of the propagation medium and the terminal equipment (antennas and transmission lines) linking the transmitter and receiver.

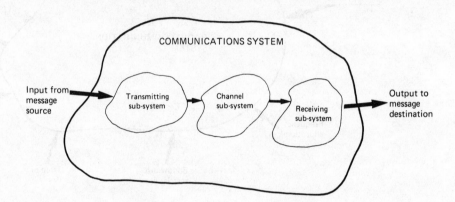

Fig. 2.5   *Decomposition of a communications system into sub-systems*

The antenna/propagation medium is a complicated interface dependent upon propagation path characteristics. It is therefore expedient to include this interface totally within the channel sub-system rather than use it as an interface between sub-systems. Using these interface definitions, the sub-systems are defined as follows:

a) The *transmitting sub-system* has the initiating information source as input and the signal into the transmission line as output.
b) The *channel sub-system* takes its input from the output of the transmitter, passes the signal across the propagation medium and delivers it to the receiver input terminal.
c) The *receiving sub-system* takes the output from the receiving transmission line and delivers the signal to its required destination.

2.2.3   Open Systems Interconnection

Before proceeding to a more detailed discussion of the three sub-systems defined above it is useful to introduce the concept of Open Systems Interconnection (OSI) which is concerned with the exchange of information between systems, allowing them to both communicate and co-operate. The ISO 7498 document[1] describes a reference model of Open Systems Interconnection.

Each connected system is considered to be logically composed of an ordered set of sub-systems. Sub-systems of the same *rank* collectively form a *layer*. Sub-systems consist of active elements called *entities*. The entities of a sub-system in one layer provide services to the layer above and use the services of the layer below. Seven layers are defined in all.

The definition of communications system shown in Figure 2.5 embraces only the first two layers:

Layer 1 (Physical Layer) which provides the mechanical, electrical, functional and procedural means to support connection and bit transmission.
Layer 2 (Data Link Layer) which accommodates techniques to achieve error-free transmission, incorporating the detection and recovery from errors which may occur in the physical layer. The Data Link Layer provides a service to Layer 3, the Network Layer.

In terms of the ISO seven-layer representation Figure 2.5 can be structured as shown in Figure 2.6. Further layers would need to be introduced if, for example, two or more communications systems were interconnected to form a network, but such a case is not considered here.

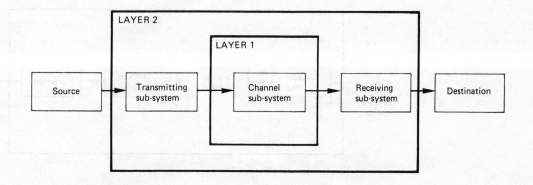

Fig. 2.6 *Open Systems Interconnection view of a communications system*

For the case of data communications, the terminology for the communications system description is often somewhat different from that described above. The data transmitter and the data receiver are often referred to by the general term of *data communication equipment* (DCE). The data source and the data sink are known by the general term of *data terminal equipment* (DTE). The communication channel with its associated terminating equipment is known as the *data link*. These definitions are illustrated in Figure 2.7. The system view of a data communications system is the conveyance of data from one DTE to another DTE via a data link.

In order to analyse the communications system of Figure 2.5 further it is now necessary to consider each constituent sub-system in more detail.

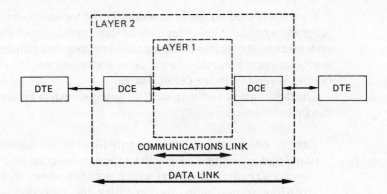

Fig. 2.7　*Generalised data communications system*

2.2.4　Transmitting Sub-system

The transmitting sub-system converts, in a series of steps shown in Figure 2.8, the message provided by the information source into a modulated electromagnetic signal that can be output to the channel sub-system.

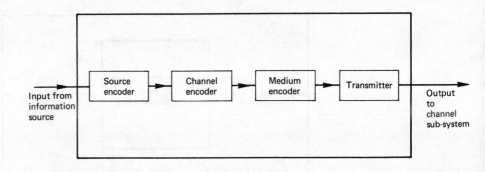

Fig. 2.8　*The transmitting sub-system*

The *source encoder* transforms the output of the information source into physical signals. These signals may, for example, take the form of voltage pulses (as from a morse key in telegraphic systems) or more continuous voltage/time functions (as from a microphone transducer in radio telephone systems).

The *channel encoder* converts the signal representing a symbol (for data systems) into another generally more complex form. The conversion process involves adding redundancy to the signals and is that part of the system which implements the encoding necessary when employing error detection or error correction.

The *medium encoder* (or modulator) operates on the encoded signals and converts them into a form suitable for transmission over the propagation medium between the transmitter and receiver. Regardless of whether the modulating signal is digital or analog, or whether amplitude, frequency, phase or another type of modulation is used, the output signal from the transmitter contains many frequency components other than the fundamental.

2.2.5 Channel Sub-system

The channel sub-system converts the signal at the output terminals of the transmitter into a roughly equivalent (but much reduced) signal at the input terminals of the receiver (see Figure 2.9).

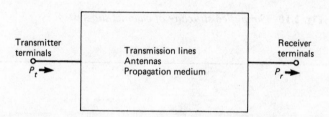

Transmitter terminals

P_t →

Transmission lines
Antennas
Propagation medium

Receiver terminals

P_r →

Fig. 2.9 *Channel sub-system as a black box demonstrating power loss*

If p_t is the power fed into the terminals of the feeder line from the transmitter and if p_r is the signal power available from the terminals of the line from the receiving antenna, then the total system power loss L, expressed in decibels, is given by

$$L = 10 \log_{10} (p_t/p_r) \tag{2.1}$$

Thus, for a given transmitter output power, as the loss L increases, the received power p_r decreases for a constant p_t. This appears to be quite a useful black box definition for the sub-system and confirms the suitability of the sub-system boundaries thus created.

In logarithmic units (2.1) can be written in the form:

$$L = P_t - P_r \tag{2.2}$$

where P_t, P_r are now expressed in decibels (dB).

The total sub-system power loss, L, is caused by three major sources:

a) Ohmic losses in the transmission (feeder) lines.
b) Ohmic and matching losses at the antenna.
c) Path losses in the propagation medium.

The ohmic losses in the antennas and their transmission lines are the same for all paths. However, the losses caused by matching the antenna to the channel depend upon the particular path over which the radio wave will propagate. There may be two or more propagation paths for a given communications link, so Figure 2.9 can be expanded as shown in Figure 2.10.

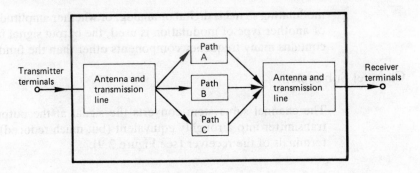

Fig. 2.10 *Simplified structure of channel sub-system*

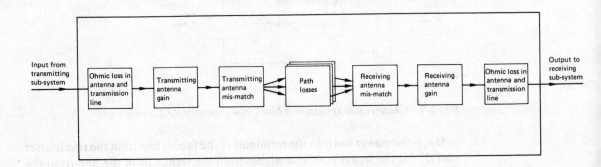

Fig. 2.11 *Loss mechanisms within the channel sub-system*

Figure 2.11 summarises the major sources of power losses. The ohmic losses in the feeder line of the transmitting antenna are caused by the resistance of the wire and the finite conductivity of the environment (induced earth currents, etc.). The rest of the power is radiated into space, mostly from the antenna, but a small amount is radiated from the transmission line. (The radiation efficiency of the antenna is given by the ratio of the power radiated to the power absorbed from the source.)

For an isotropic radiator the power is radiated equally in all directions and the power radiated in any direction can be taken as a reference. Actual antennas have radiation patterns for which the power flux is maintained in certain preferred directions and this must be balanced by a loss in other directions. Depending upon the polarisation characteristics of the antenna and the mode of propagation there may be power losses caused by polarisation mismatch (see Figure 2.11).

The receiving antenna picks up only an infinitesimal amount of the power radiated as well as unwanted radiations from many interfering sources. In practice, the total signal arriving at the antenna may be composed of individual signals which have traversed different paths and which have differing characteristic polarisations.

The presence of additional paths increases the received power and, therefore, diminishes the effective loss of the system. The ohmic losses in the antennas and their transmission lines are the same for all paths (see Figure 2.11). However, the effective antenna gain depends on the particular propagation path. (The effective antenna gain is a combination of the antenna gain and polarisation mis-match losses for the particular angle of take-off and wave polarisation.)

2.2.6 Receiving Sub-system

The receiving sub-system (Figure 2.12) takes the message from the channel sub-system and processes it into a form suitable for the destination to receive.

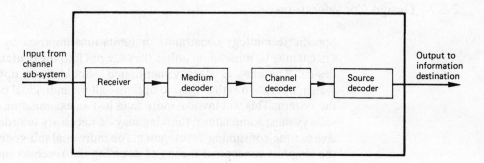

Fig. 2.12 *The receiving sub-system*

The amplitude of a composite wave signal at the receiver is given by the vector sum of its components; thus, if the components are in phase, the resultant amplitude will be the algebraic sum of the individual wave amplitudes. If the phases of the separate components are varying randomly with respect to each other, the resultant signal power, averaged over many fading cycles, is given by the sum of the powers in the component waves.

The received (wanted) signal at the antenna will be combined with unwanted signals from a variety of sources. There are many sources of radio frequency interference and noise. Some are in a sense controllable by careful choice of antenna design and siting, whilst others are likely to be present in all cases. The receiver itself contributes some thermal noise. Indeed, noise is a fundamental concept that underlies many facets of receiver performance. It is linked to a range of other receiver parameters, one of the most important of which is *sensitivity*. (This is expressed as the smallest signal required to give a specified signal-to-noise ratio for a particular receiver bandwidth.) At the receiver output the signal carrying the desired information must be of sufficient power relative to all of the unwanted signals and noise combined to provide for a faithful reproduction of the original message.

The *medium decoder* (or detector) operates inversely to the medium encoder (modulator). It converts the modulated signals that are received by the receiver into signals similar to those at the output of the channel encoder. The device often acts as a primary decision maker and, in a binary system, may decide whether the received pulse is a binary 1 or a binary 0. The output signals from the medium decoder are used in the remaining decoding part of the receiver.

The *channel decoder* operates inversely to the channel encoder. It converts a coded signal which, as a result of transmission impairments, may differ from that at the output of the channel encoder. A signal is produced which, ideally, should correspond directly to the signal originally output from the source encoder (see Figure 2.8).

The *source decoder* operates inversely to the source encoder. It converts physical signals into symbols suitable for use by the destination. For example the signal may be converted to the output from a telephone headset or from a teleprinter. The signals which constitute the input to the source decoder depend upon any previous decisions made at the medium decoder.

2.2.7   Design Considerations

Specific technology constraints or limitations imposed by a particular sub-system may be misleading unless they are put into the context of a total system design. The aim, as already mentioned, should be to optimise the overall system operation rather than to optimise all the individual components within the system. This will involve more than just an examination of the constituent sub-systems; some lateral thinking may be necessary in order to avoid expensive or time-consuming developments on individual sub-systems. For example, the simplest conceptual means of doubling the received signal strength is to double the transmitter output power. This may be costly, requiring special development or enhancements; the result may introduce spurious effects at the transmit terminal, and it will cause more 'pollution' of the radio spectrum.

From the systems viewpoint the effect of doubling the received signal strength could be achieved by halving the received noise, since it is signal-to-noise ratio that determines grade of service. The noise incident at the receive antenna might be reduced by the use of a suitably oriented directional antenna. Thus the reason for a particular specification of transmitter output power may originally have been to achieve a desired grade of service which could be established more easily and more cheaply by consideration of receive antenna characteristics. Solutions to system problems which are ideal for fixed point-to-point links may be wholly inappropriate for use in a mobile communications system such as an air-to-ground link. Then some other means for improving system performance might be available, for example choice of a better frequency or message relay via another aircraft.

With the many powerful unwanted signals that are present in the HF band, the design parameters of the HF receiver require careful consideration. The system designer must be aware how the interference environment can impact upon system performance, not only in a direct way by reducing the signal-to-noise ratio at the antenna but also in an indirect way within the receiver. For example, *intermodulation products* occur when two large unwanted signals beat together in a non-linear receiver stage to give a product at the wanted frequency. A noisy product can also be produced at the wanted frequency by large unwanted signals mixing with the noise sidebands of the local oscillator, a phenomenon known as *reciprocal mixing*.

A thorough appreciation of the use for which the communications system is intended must therefore be available to the designers from the very earliest stage. The communications system may well be a sub-system within some larger information management system and this, too, must have its requirements defined comprehensively and unambiguously. Communications are fundamental to information-based systems be they manual or automatic. Central to the requirements of the user of future systems is the need to communicate efficiently, easily, reliably, unambiguously and, within the military scenario, securely.

Communications are now recognised as a vital and integral element within C^3I systems (that is why C^2 systems became C^3 systems and then C^3I systems). It is therefore important that the subject of HF communications be addressed from the systems viewpoint rather than from details of individual hardware design aspects such as the transmitter or receiver circuitry. No longer are communications systems the exclusive domain of communications engineers. Today the systems analyst and designer of computer-based systems must have a considerable appreciation of the technical principles of the wide range of communications techniques.

The resurgence of interest in the HF band has meant that many new systems being proposed are incorporating HF links. HF is a vital ingredient in the strategic mix of communications media necessary to maintain adequate C^3 under all conditions. Technology provides the ability to improve system performance, reliability and availability beyond that which was possible only a few years ago except with the assistance of a dedicated and experienced operator. The unique problems associated with HF make it essential that today's systems designer has a fundamental appreciation of HF communications.

2.3 HF Communications Systems

2.3.1 Propagation Characteristics

Propagation in the HF band, except at short ranges (typically less than 150 km), depends upon the reflection of radio waves from ionised layers in the Earth's atmosphere. A low transmission power can often achieve extremely long range when the proper transmission frequency is used. A number of factors govern the state of the propagation path and the choice of operating frequency, such as the time of day, the season, the level of solar activity, the path length and its orientation.

For long-range links the *sky wave* leaves the HF transmitting antenna at an inclined angle and travels upwards to be reflected from the ionosphere. After the radio signal has left the transmitting antenna it may lose energy density by several processes, such as:

a) The spatial spreading of the energy (inverse distance squared dependence).
b) Polarisation mis-matching at the transmit antenna.
c) Absorption of energy within the ionosphere due to electron collisions. (The energy in the wave is converted into heat.)
d) Scattering caused by ionospheric irregularities.
e) Polarisation mis-match at the receive antenna. (For certain directions there may be a gain due to the antenna pattern; in other directions there will be a loss.)

Radio frequency energy from the transmitted signal also remains near the Earth's surface; this is termed the *ground wave*. Depending upon terrain conditions and the transmitter's output power, the ground wave can be used for communicating over distances up to 100 km. Ranges of about 160 km can sometimes be obtained at transmission frequencies below 4 MHz but, as the operating frequency is increased, the signal experiences excessive attenuation. Ground wave signals can be received at greater ranges over sea water than over average soil or dry rocky terrain. (A more detailed examination of ground wave propagation is given in Chapter 3.) Ground wave communications are more straightforward than sky wave; it can be assumed that the ground wave is merely an attenuated, delayed but otherwise undistorted version of the transmitted signal. Ionospheric sky wave returns, however, in addition to experiencing a much greater variability of attenuation and delay, also suffer from fading, frequency or Doppler shifting, spreading, time dispersion and delay distortion. These features are discussed in detail in Chapter 4.

Three distinct ranges can be identified for HF propagation. These depend upon the relative importance of the ground wave and the sky wave.

a) At *short ranges* with vertical polarisation, ground diffraction predominates. With horizontal polarisation there would be no appreciable ground wave.

b) At *medium and long ranges* with vertical polarisation (and at every range for horizontal polarisation) the ionospheric wave, if present, is the predominant mode.

c) At *intermediate ranges* there can be observed either:

(i) Co-existence of ground and sky wave modes (see Figure 2.13a) with very strong fading if their mean values are identical; or

(ii) A silent zone (see Figure 2.13b) because the ground wave has become negligible and the sky wave has not yet appeared.

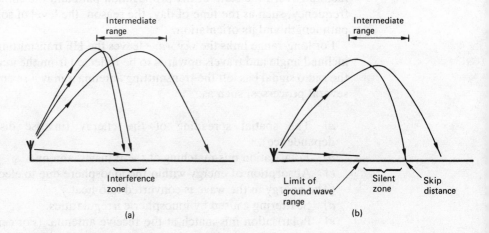

Fig. 2.13 *Range characteristics for HF propagation*

CHARACT-ERISTIC	GROUND WAVE	SHORT-RANGE SKY WAVE
Propagation	Power decreases with distance. Doubling range reduces field strength by approx. 12 dB over land paths. Depends upon ground electrical characteristics, topology and vegetation.	Essentially constant received power over a circle radius 300 km centred on transmitter. Independent of ground characteristics.
Frequency band	All frequencies in the range 2–30 MHz could be used. Least attenuation at lower frequencies.	Narrow window only is available at any one time. Window highly dependent upon prevailing conditions.
Antenna	Vertically polarised antennas required for efficient ground wave generation.	Good high angle coverage must be provided by the antenna.
Polarisation	Vertical polarisation must be used. Horizontal polarisation very heavily attenuated.	Horizontal polarisation preferable to avoid ground wave interference.
Performance	Can be severely range limited by terrain. Performance highly dependent upon range.	Essentially uniform coverage over a wide area. Performance heavily dependent upon correct choice of frequency.

Table 2.1 Comparison of ground wave and short-range sky wave characteristics

The existence of condition c (i) or c (ii) depends upon the frequency of the signal and the prevailing ionospheric conditions. Condition c (i) is more likely to occur at the low end of the HF band, whilst c (ii) occurs for higher frequencies when there is a *skip distance*, which is the minimum achievable range by an ionospherically reflected signal.

When conditions of c (i) pertain, the sky wave is known as 'nearly vertically incident' since to achieve the short ranges where the ground wave is still present the ionospherically reflected ray must be incident upon the ionosphere at a very steep angle. A comparison of ground wave and sky wave for conditions of c (i) is given in Table 2.1.

2.3.2 A Unique Role for HF

Dependence upon the sky wave for long-distance communications suffers a number of disadvantages, the main ones being:

a) The variability of propagation conditions which, for optimum results, requires frequent changes in the operating frequency.

b) The interruption of communications by ionospheric storms.

c) The large number of possible propagation paths resulting in the time dispersion of a single signal.

d) The large and rapid phase fluctuations.

e) The high levels of interference.

f) The frequency distortion suffered by wideband signals.

At certain times it may be impossible to obtain a satisfactory propagation path at all, for periods (depending upon path geometry) ranging from a few minutes to several hours. The operational use of HF must, therefore, take into account the statistics of the availability of a given link before relying upon HF in a particular role, although it is often the case that no alternative communications path can be provided.

In spite of these shortcomings the HF spectrum from 2 to 30 MHz still offers major advantages over all other frequencies. The HF radio transmission medium has the characteristic of being able to provide beyond-line-of-sight communications over ranges up to thousands of kilometres without the need for repeaters. High frequency radio waves have been, and continue to be, one of the basic vehicles for long distance transmission of information. The reasons for this may be summarised as follows:

a) Low cost of terminal equipment

b) Low power requirements

c) Adequate bandwidths

d) Adequate signal strengths.

While HF links cannot handle traffic densities as high as other means of communication, they can operate in the voice, data, teleprinter and facsimile modes.

Many radio frequency bands, ranging from ELF up to VHF, can be reflected from the atmosphere. Why, then, is HF the primary sky wave band? The answer lies in Figure 2.14. Atmospheric noise, both natural and man-made, increases with decreasing frequency. Simultaneously, antenna efficiency decreases exponentially as the antenna is shortened in terms of wavelength. An efficient ELF antenna of one-quarter wavelength would be 2500 km in length! The optimum compromise between low atmospheric noise, practical antenna efficiency and communications range occurs in the HF band as shown by the shaded region in Figure 2.14.

Medium wave frequencies (300 kHz to 2 MHz) suffer very heavy absorption during the day, while on lower frequencies there is insufficient bandwidth for more than a few voice channels. Frequencies above 30 MHz are not, normally, reflected from the ionosphere. VHF can be reflected up to 1800 km or less, but only under rare sporadic circumstances. VHF meteor reflections are reliable for low rate data but have approximately the same 1800 km range limitation.

Frequencies in the VHF band and above are primarily limited to line-of-sight communications. For all practical purposes UHF can only be operated successfully when there is visual contact. VHF, though better than UHF in this respect, also suffers from screening difficulties. HF signals suffer much less screening from physical obstacles, while the attenuating effect of snow, fog and rain is very much less than at higher frequencies. Long-range systems using VHF or tropospheric-scatter techniques use highly directional, narrowband antennas. In a tropospheric-scatter link these are expensive, large and cumbersome and require accurate siting.

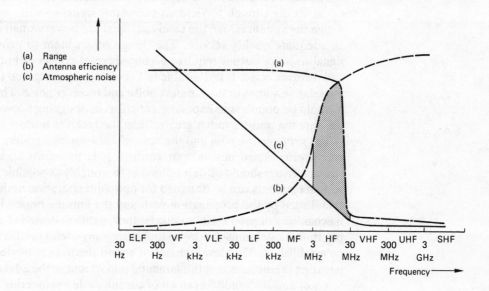

(a) Range
(b) Antenna efficiency
(c) Atmospheric noise

Fig. 2.14 *Radio characteristics by frequency band*

HF equipment is generally of low cost and can be located in fixed or mobile installations. HF antennas can be either omnidirectional or directional, wideband or designed to operate at a specific frequency, depending on the requirement.

HF is ideally suited for use in peacetime by widely dispersed military units which need long-range communications paths, as well as by the same units in wartime even if a smaller area of control is assumed. HF antennas and equipment can be deployed rapidly to provide immediate military command-post communications without the need for careful siting, as is required for line-of-sight transmission.

The HF band is therefore the frequency band with a compromise of sky wave features:

a) Highest frequency band that will reflect from the ionosphere to provide long-range communications.
b) Far more variable reflection and therefore less predictable than for lower frequency bands.
c) Smaller antennas required than for lower frequencies and thus greater efficiencies achievable.
d) Lower noise levels than lower frequency bands.

2.3.3   System Design Overview

The major constraint for a satellite link is to maintain an adequate signal-to-noise ratio for the high data rates that are required. In contrast, constraints on HF links centre around the dispersive characteristics of the transmission medium and in the high levels of interference that may be encountered.

The basic problem in the design of a communications system is to receive a signal strong enough to yield an *acceptable signal-to-noise ratio*. If the noise within the system is low, the received signal can be very small and still produce an adequate quality service. The design of a system to provide satisfactory signal-to-noise output requires a knowledge of effects of both antenna noise and receiver noise. It is desirable to know, over the required frequency range, the relative values of the antenna noise and receiver noise. Thus, for example, it would be pointless to expend great effort in designing a low-noise receiver if the antenna noise is much greater than the receiver noise.

The *propagation path* and the *natural noise* are two system elements whose characteristics are beyond our control. It is therefore apparent that these characteristics should be determined as thoroughly as possible so that the other system elements can be designed for optimum operation under the conditions established by the propagation path and the limiting noise. The fundamental mechanisms of propagation must be thoroughly understood and suitable propagation models must be postulated before any useful prediction formulae can be established. The development of sound theoretical models for radio propagation predictions is of fundamental importance in the advancement of radio science, and such models can be of considerable engineering value. Measurements of radio wave propagation are usually time consuming and require expensive equipment and widespread facilities. It would require an enormous measurement programme to obtain adequate statistical samples of all the host of variables which influence the propagation medium formed by the Earth's surface and surrounding atmosphere. The results deduced from the theoretical models developed should, of course, be measured against experimental results as far as is possible to check their validity. It is necessary to analyse critically the simplified assumptions of these models to indicate their accuracy and to pinpoint where improvements can be made.

The amount of *transmitted energy* that arrives at the receiver over a propagation path depends on many factors. One of the most important is the mode or combination of modes of propagation which predominate. When system performance is to be estimated it is important to know what mechanism will be useful so that reliable estimates of system loss and its variation can be made. For example, severe fading may be encountered at ranges where the amplitude of the ground and ionospherically propagated waves are of the same order of magnitude. The phase path of the ionospheric wave changes continually so that it is alternatively in and out of phase with the ground wave. When it is of near equal amplitude and of opposite phase, deep fades occur. An estimate of temporal and spatial variation of these fades is desirable.

The design of HF systems depends upon accurate predictions and new technology to improve circuit reliability. System planners need to know what *frequency ranges* their systems should be capable of covering, what *transmitter powers* are necessary to overcome the background noise at the receiver, and what *antenna configurations* would be most suited to the applications required. These are all factors dictated by propagation considerations and long-term prediction techniques are available for this purpose.

There are a number of factors which affect the final choice of *modulation scheme* used for any communications band or medium. For satellite and microwave communications, the prime concern is normally bandwidth and power efficiency and the performance with limiting and filtering. However, for HF communications, these are of less concern than the performance and robustness of the modulation scheme which must overcome the variable characteristics of the HF channel. It is often not only the modulation scheme itself which is important when comparing performance under fading and multipath conditions but also the associated coding and equalisation processes. These parameters are examined in Chapter 9.

A given communications system can only be as effective as the weakest link in the chain from transmitter to receiver. For example, in the case of air-ground systems the ground terminal is just as important as the aircraft terminal and must be specified accordingly. Since there are likely to be many more aircraft than ground stations, and all airborne electronic equipment penalises the aircraft performance because of its mass, volume and power consumption, it is important to keep the complexity of the airborne communications terminal as low as possible.

It may be expensive to realise communications reliability targets; cost-effectiveness studies must be carried out. The main factors which degrade reliability need to be identified. Assessments can then be made of reliability improvement which may be achieved by realistic modifications to the total system.

2.3.4   The Power Level Diagram

In the initial stages of planning it is important to have an overall picture of the HF link. One method of achieving this is with a *power level diagram*. The advantage of the power level diagram is that it can show at a glance the major loss factors within a given communication link. It has a vertical axis (see Figures 2.15 – 2.17) scaled in power usually in dB relative to 1 watt (dBW) or in dB relative to 1 milliwatt (dBm), thus 30 dBm is equivalent to 0 dBW. The horizontal axis is arbitrarily scaled but progresses from left to right from the transmitter output power P_T via the propagation path to the receiver input. Gains are indicated by positive slopes, losses by negative slopes. The space between the broken lines drawn parallel to the horizontal axis represents signal-to-noise ratio prior to the receiver demodulator.

Consider the example of Figure 2.15 which shows a 5 MHz ground wave link at a range of 100 km from a 1 kW transmitter in the presence of atmospheric noise at two different times of year. The sky wave signal is ignored. The signal-to-noise ratio is only 4 dB in a 3 kHz bandwidth (35 dBHz) on a summer evening when noise levels are high. This is not adequate for a voice circuit. The same link on a winter morning has a signal-to-noise ratio of 30 dB, a good-quality circuit.

Figure 2.16 shows a point-to-point sky wave link for 4 and 15 MHz at a range of 2000 km at noon. Note how much more strongly the 4 MHz signal is

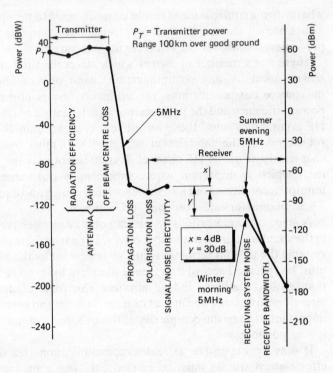

Fig. 2.15 *Example of a power level diagram for a short-range ground wave link*

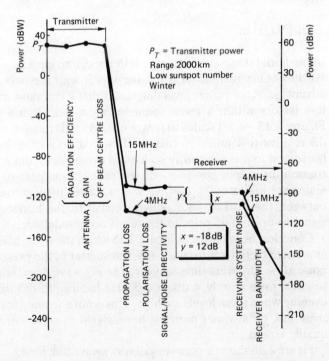

Fig. 2.16 *Example of a daytime (mid-day) long-range sky wave point-to-point link*

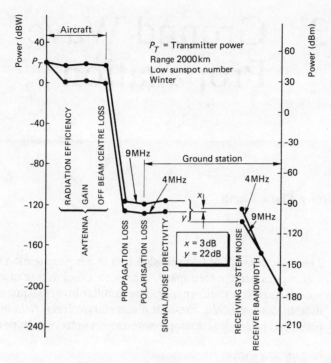

Fig. 2.17 *Example of a night-time (midnight) long-range sky wave air-to-ground link*

attenuated than is the 15 MHz channel. The respective signal-to-noise ratios are −18 dB and 12 dB. This clearly emphasises the importance of selecting the correct frequency.

An example of a night-time air-ground link over the same range as the previous example is shown in Figure 2.17. Although the propagation loss is less for 4 MHz than for 9 MHz the overall signal power is reduced due to the poorer aircraft antenna efficiency for 4 MHz. In this case 9 MHz (signal-to-noise ratio of 22 dB) would provide an adequate channel, whilst 4 MHz (3 dB) would be unworkable.

These examples serve to show for the evaluation of HF circuit performance the importance not only of the propagation path loss and noise levels but also of the choice of correct frequency.

Before proceeding further with a more detailed system performance assessment it is necessary to examine the fundamental physical constraints on the link caused by the propagation effects, both for ground waves and sky waves, and the ambient noise levels. These are described in the following three chapters. The discussion of system performance assessment resumes in Chapter 6.

3 Ground Wave Propagation

3.1 Propagation over a Plane Earth

3.1.1 Free Space Propagation

The basic concept of transmission loss in the propagation of radio waves is the loss experienced in free space, that is, in a region free of all objects that might absorb or reflect radio energy. The familiar inverse square law in optics can be used to determine E_0, the root mean square (rms) *field intensity in free space*, produced by an ideal isotropic antenna. In units of volts per metre E_0 is given by

$$E_0 = (30P)^{1/2}/d \tag{3.1}$$

where d is the range in metres and P the transmitted power in watts. The factor $30^{1/2}$ becomes $45^{1/2}$ when the isotropic antenna is replaced by a very short vertical dipole.

Free space transmission assumes that the atmosphere is perfectly uniform and non-absorbing and that the Earth is either infinitely distant or that its reflection coefficient is negligible. The presence of the ground modifies both the generation and propagation of radio waves so that the received power or field intensity is ordinarily less than would be found in free space.

For an antenna at a short distance above a perfectly conducting ground, the power is radiated only into one hemisphere. Then the power density in a given direction is doubled and the field intensity is increased by a factor of $2^{1/2}$. Thus, for a very short vertical dipole near the ground, the rms field strength becomes

$$E_0 = 300P_k^{1/2}/d \tag{3.2}$$

with P_k in kilowatts and d in metres. Equation (3.2) has been adopted as an international standard by the International Radio Consultative Committee (CCIR) at their 1980 meeting[1].

3.1.2 Electrical Characteristics of the Ground

The electrical characteristics of the ground can be expressed by three constants, namely the relative permeability, the dielectric constant and the conductivity. The *relative permeability* can normally be regarded as unity, so that in most propagation problems only the *dielectric constant* ϵ and the *conductivity* σ are of concern. These two parameters jointly influence wave propagation in accordance with the following expression for a complex dielectric constant relative to a vacuum:

$$\epsilon' = \epsilon - 60i\sigma\lambda \tag{3.3}$$

where σ is in siemens per metre and λ is the free space wavelength in metres. The relative importance of the displacement and conduction current densities is given by the ratio of ϵ to $60\sigma\lambda$ for the appropriate wavelength.

The effect of the electrical characteristics of the ground upon the propagation of the radio waves is then given by[2]

$$E/E_0 = \underset{\substack{\text{Direct} \\ \text{Wave}}}{1} + \underset{\substack{\text{Reflected} \\ \text{Wave}}}{Re^{i\Delta}} + \underset{\substack{\text{Surface} \\ \text{Wave}}}{(1-R)Ae^{i\Delta}} + \underset{\substack{\text{Induction} \\ \text{field and} \\ \text{secondary effects}}}{\dots} \tag{3.4}$$

where R, A and Δ are all functions of ϵ' and

R = complex *reflection coefficient* of the ground for the wave polarisation of interest.

A = surface wave *attenuation factor*.

Δ = phase difference caused by the path difference between the direct and ground reflected waves.

3.1.3 Ground Wave Components

The behaviour of equation (3.4) can be demonstrated most easily by considering two limiting cases. These enable the ground wave to be considered either purely as a surface wave or purely as a space wave.

a) For antennas approaching ground level, Δ approaches zero and R approaches -1. Then the first two terms of equation (3.4) cancel, leaving the surface wave. The surface wave term arises because the Earth is not a perfect reflector, causing some energy to be transmitted into the ground.

b) For transmitting and receiving antennas which are both elevated by more than a few wavelengths, the surface wave can be neglected. The ground wave is then known as the space wave. For near grazing paths, R approaches -1 and equation (3.4) becomes

$$|E/E_0| = 2 \sin (\tfrac{1}{2}\Delta) \tag{3.5}$$

This expression is the sum of the direct and ground reflected waves.

For the wavelengths used in the HF band it is the surface wave that predominates in the expression for the ground wave, since the antennas are unlikely to be elevated by more than a few wavelengths (except in the case of aircraft to aircraft links).

3.2   The Surface Wave

3.2.1   General Principles

The phenomenon of diffraction occurs for all types of wave motion; it causes bending of the wave around any obstacle which it passes. For the surface wave the obstacle is the Earth and the amount of diffraction depends upon the ratio of the wavelength to the radius of the Earth. The degree of diffraction decreases steadily as the wavelength is decreased because the ground is an imperfect conductor. Energy is absorbed by currents induced in the Earth; there is a continuous flow of energy from the wave downwards into the Earth. The wave front is therefore tilted slightly forward, as shown in Figure 3.1 and hence the bending of the wave is assisted.

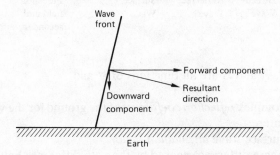

Fig. 3.1   *Tilting of wave front caused by ground losses*

Consider the case of transmitting and receiving antennas both situated on the surface of the Earth. The *radiated field* produced at a distance d from the transmitter may be expressed as

$$E = KFP^{1/2}/d \qquad (3.6)$$

where

P is the total radiated power
K is a constant which depends upon the antenna characteristics
F is the *attenuation factor*, less than or equal to unity.

The factor F depends upon wave frequency, ground characteristics, wave polarisation and distance. The first three parameters influence F as follows:

a) F decreases (i.e. attenuation increases) with increasing wave frequency;
b) F decreases with decreasing ground conductivity;
c) F is much smaller for horizontal than vertical wave polarisation.

3.2.2   Zonal Relationships

The dependence of F upon distance from the transmitter is rather complicated. It can best be explained by a consideration of 'zonal' relationships. As distance from the transmitter is increased from zero, the following zones are found:

1 Direct radiation zone
2 Sommerfeld zone
3 Diffraction zone.

Each has its own individual effect upon the attenuation and a schematic representation of range dependence is shown in Figure 3.2.

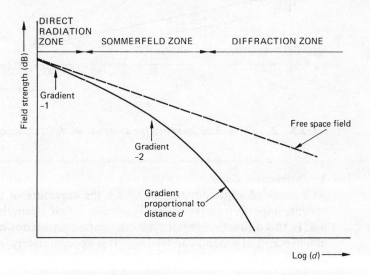

Fig. 3.2 *Zonal relationships for antennas on the ground*

1 Direct Radiation Zone

At small distances from the transmitter, the radio waves travel as in free space and F is equal to unity. The radiated field E, see equation (3.6), then varies as $1/d$ and is calculated in the same manner as for geometric optics.

2 Sommerfeld Zone

As the distance increases, the field near the surface due to a source on the surface can be described by the Sommerfeld Flat Earth theory[3]. As the range further increases, corrections due to the effects of the Earth's curvature can be applied[4]. Typically, in the Sommerfeld zone, F becomes proportional to $1/d$, so that the field E becomes proportional to $1/d^2$.

The extended Sommerfeld theory is useful[5] out to distances of the order $10\lambda^{1/3}$ km where λ is the wavelength in metres. However, the boundary between the direct radiation zone and the Sommerfeld zone depends upon the nature of the ground (see Figure 3.3). Note that for sufficiently low frequencies, the Sommerfeld zone does not exist; there is a direct transition from the direct radiation zone to the diffraction zone. This transition depends upon the nature of the ground.

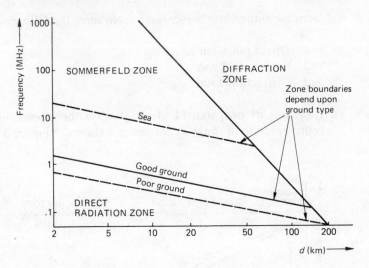

Fig. 3.3   *Zone boundary locations for antennas on different types of ground*

3   *Diffraction Zone*

At a range of approximately 10 $\lambda^{1/3}$ km the curvature of the Earth begins to become important and the overall decrease in field strength becomes exponential. In this diffraction region, the attenuation settles down to a value which is independent of ground conductivity. It is approximately $0.62/\lambda^{1/3}$ dB/km (λ in metres).

3.2.3   Effect of Antenna Height

When the antennas are close to the Earth the surface wave predominates, but the presence of the two components of the space wave – the direct and reflected waves – must not be forgotten. At its point of reflection the reflected wave suffers a change of phase which, for small angles of incidence, is very nearly 180°. Thus when the antennas are close to the surface the two components of the space wave almost cancel.

As antenna heights are increased, the path length of the reflected ray becomes somewhat greater than that of the direct ray. In this way a further phase change is introduced and the two components no longer cancel. For small heights the change in field strength is small. In some cases the relative phase of the surface wave and the space wave results in a slight decrease in field strength at first, but eventually the signal increases steadily with height.

When heights of the antennas are raised above the surface of the Earth then equation (3.6) must be modified as follows:

$$E = KFP^{1/2} H(h_1) H(h_2)/d \tag{3.7}$$

where $H(h)$ is the height gain factor for an antenna at height h above the surface and h_1, h_2 are the heights of the transmitting and receiving antennas. H is generally (but not always) greater than unity.

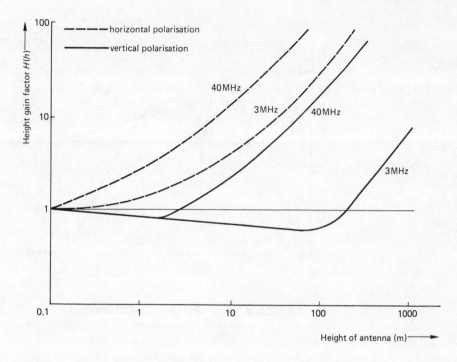

Fig. 3.4 *Height gain factors over good ground*

As the height of the antenna is increased, H may be equal to or even slightly smaller than unity (see Figure 3.4) until a boundary height is reached. Above this height, the factor H increases in proportion to the height of the antenna until the region is reached (at about $35\lambda^{2/3}$ m height, λ in metres) where geometric options can be used.

The effect of antenna height above the surface is more pronounced over poor ground than over sea water. Height gain has more effect upon horizontal polarisation than upon vertical polarisation. When both transmitting and receiving antenna heights become three wavelengths or more, the wave polarisation no longer has an important effect.

3.3 The Space Wave

3.3.1 Zonal Relationships

When the transmitting and receiving antennas are raised above the ground by more than $35\lambda^{2/3}$, the effect of the Earth's surface becomes small, and the space wave predominates. For the space wave, the following zones are encountered as distance from the transmitter is increased:

1 Interference (or line of sight) zone
2 Radio horizon zone
3 Diffraction zone.

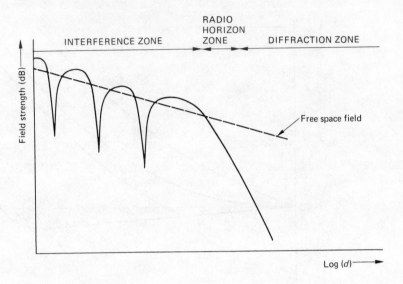

Fig. 3.5 *Zonal relationships for elevated antennas*

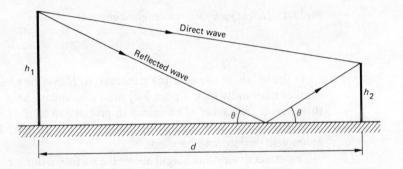

Fig. 3.6 *Direct and reflected wave components of the space wave*

A schematic representation of range dependence is shown in Figure 3.5.

1 *Interference Zone*

In this zone, the total field is composed of the sum of the direct and reflected waves (see Figure 3.6). The field strength E_0 of the direct wave at a distance d is given by geometric optics as in equation (3.1). The field strength of the reflected wave differs from that of the direct wave in four main respects:

a) The propagation distance to the receiver is longer. This means that the field strength is reduced and there is a phase difference between the direct and reflected wave.

b) The gain of the transmitting antenna may be different for the two component waves since their directions of propagation are different.

c) The reflected wave is modified by a complex *reflection coefficient R*, which depends upon the angle of incidence of the wave, its polarisation and the ground characteristics. Reflection from a lossy ground produces changes in both magnitude and phase of the reflected signal.

d) Reflection from the Earth's spherical surface produces a divergence of ray paths. This has the effect of further reducing the reflected wave strength. The ratio of the field strength after reflection from a spherical surface to that after reflection from a plane surface is known as the *divergence coefficient D*.

Taking into account these differences, the field strength of the reflected wave E_R compared to the direct wave E_0 then becomes

$$E_R = (G_R/G_D)^{1/2} (d_D/d_R) \, D \, R \, E_0 = \alpha E_0 \tag{3.8}$$

where G_R, G_D are the antenna gains and d_R, d_D the total path lengths for the reflected and direct waves respectively. When the phase factor due to path difference is included, the total field then becomes

$$E = E_R + E_0 = E_0 \left[1 + \alpha \exp \left(-2\pi i (d_R - d_D)/\lambda \right) \right] \tag{3.9}$$

In practice the situation is rather more complicated than (3.9) shows since the radio ray trajectories may not be linear. Tropospheric refraction has the effect of bending the ray paths, further complicating the computation.

2   Radio Horizon Zone

Reflection theory in geometric optics is valid only when diffraction has a minor effect, i.e. for cases when the waves are not too grazing in incidence.

Suppose the grazing angle is less than

$$\psi = \lambda / 2\pi m a \tag{3.10}$$

where a = Earth's radius

m = parameter between unity and 4/3 used to modify Earth's radius resulting from tropospheric effects.

Under these conditions ray theory becomes invalid and the surface wave may contribute appreciably to the total field strength. In this zone, the field strength must be fitted to that in the interference zone and the diffraction zone. The exact calculation of this fit is rather difficult[6], since absorption and diffraction are combined in a complex fashion. The total attenuation is obtained by a series expansion which has terms that are tedious to calculate and converge slowly.

3   Diffraction Zone

Beyond the radio horizon, the diffraction zone is entered. Here the attenuation settles to the value $0.62/\lambda^{1/3}$ dB/km as mentioned on page 38.

3.3.2   Effect of Antenna Height

Consider the case for which the reflection angle θ (see Figure 3.6) is greater than the limiting angle ψ defined in equation (3.10), but is small enough to ensure a reflection coefficient of approximately -1. The difference in path length between the reflected wave and the direct wave is approximately $2h_1h_2/d$ when both h_1 and h_2 are much less than d. There is a series of maxima equal to twice the field in free space (see equation (3.5)) when

$$h_1h_2/\lambda d = (2n + 1)/4 \qquad (n = 0,1,2,...)$$

and a series of minima when   (3.11)

$$h_1h_2/\lambda d = \tfrac{1}{2}n \qquad (n = 1,2,3...)$$

where n is an integer.

When the angle θ is increased the effects produced are different depending upon wave polarisation.

a) For *horizontal polarisation*, the modulus of the reflection coefficient decreases and the minima are less pronounced.

b) For *vertical polarisation*, the effects become complicated because both the phase angle and the modulus of the reflection coefficient change rapidly with changes in θ.

3.4   Deviations from Simplified Model

3.4.1   General Considerations

The theory outlined above must be modified to account for a number of other physical effects and influences that are imposed upon the radio wave as it propagates over the surface of the Earth.

The extent to which the lower strata influence the effective Earth constants depends upon the depth of penetration of the radio energy. The *penetration depth* δ is defined as that depth at which the wave has been attenuated to $1/e$ (or 37%) of its value at the surface. The penetration depth for a frequency of 10 MHz is shown for different types of ground in Table 3.1. For frequencies in the HF band only the surface of the ground usually needs to be considered since δ, even for poor conductivity soil, is no more than a few metres; for sea water it is much less than a metre.

The radio energy received at a point travels not only by the direct path from the transmitter but also by a large number of indirect paths distributed on either side of the direct path. It is necessary, therefore, to consider the constants of the ground over the area covered by the lateral spread of the wave paths as well as along the direct path itself. The most important region is the first Fresnel (half-wave) zone. This is the ellipse having the transmitter and receiver positions as its foci and axes of $(d + \lambda/2)$ and $(d\lambda)^{1/2}$ respectively, where d is the length of the direct path and λ is the wavelength.

Just beyond the radio horizon of a transmitting antenna the observed signal strength results from a variety of propagation mechanisms. These may include diffraction over ridges and hills as well as diffraction by the Earth's curvature. At one extreme, the case of diffraction over high, isolated obstacles, knife-edge diffraction theory gives theoretical results that agree fairly well with observations[7]. Field strengths are similar to near free-space values, showing a low rate of attenuation with distance. At the other extreme is diffraction over a smooth spherical Earth. This condition results in low field strengths which are soon exceeded, beyond the horizon of a transmitting antenna, by radio fields produced by reflection from elevated layers or by forward-scatter radio waves.

TERRAIN TYPE	PENETRATION DEPTH (m)
Sea Water	0.1
Wet Ground	3
Fresh Water	10
Medium Dry Ground	15

Table 3.1 Penetration depth of 10 MHz wave as a function of terrain

Theoretical methods[7] have been developed to handle certain idealisations of terrain features, such as bluffs, cliffs and knife-edge obstacles in a transmission path. However, in most cases of propagation over land it is extremely difficult to take into account the roughness and irregularities of terrain features and environmental clutter such as vegetation, buildings, bridges and electric power lines.

It is not possible to make simple general statements regarding the influence of the terrain and the vegetation on propagation. It is a complex function of frequency, ground constants, tropospheric variations, path geometry, season and vegetation density.

3.4.2 Ground Conductivity

A number of physical factors influence the effective conductivity of the ground:

1 *Moisture Content*
The moisture content of the ground is a major factor in determining its electrical characteristics. The moisture of a particular soil may vary considerably from one site to another, due to differences in the general geological formation which will affect drainage. At depths of one metre or more, the 'wetness' of the soil at a particular site tends to be substantially constant all the year round. It may of course increase during rain, but the drainage of the soil and surface evaporation soon reduce it to its normal value after the rain has stopped.

2 *Temperature*
At low frequencies the temperature coefficient of conductivity is of the order of 2% per degree Celsius while that of the dielectric constant is negligible. At the freezing point of water there is generally a large decrease in both constants. However, since the range of temperature variation during the year decreases

rapidly with depth, temperature effects are likely to be important only at high frequencies for which the penetration of the waves is small or when the ground is frozen to a considerable depth.

3   *General Geological Structure*
The ground over which a radio wave propagates is not usually homogeneous; the effective ground constants will often be determined by several different types of soil. The effective constants over an area or along a path are determined, not only by the nature of the surface soils, but also by the nature of the underlying strata. These lower strata may have a direct effect in that they form part of the medium through which the waves travel or they may have an indirect effect by determining the water level in the upper strata.

4   *Energy Absorption by Surface Objects*
Although surface objects have no direct influence on the constants of the ground itself, they can contribute appreciably to the attenuation of ground waves. Particularly high attenuation rates[8] are associated with transmission loss in wooded terrain at frequencies above about 30 MHz. Such attenuation may increase even more when the trees are covered with wet snow, and under conditions of rain when the trees are in leaf.

3.4.3   Terrain Irregularities

Most irregularities in a terrain profile are not isolated obstacles[9] that are free of the influence of nearby hills and valleys. To a first approximation such irregularities can be characterised by the difference, Δh, of terrain heights exceeded for 10% and 90% of the propagation path.

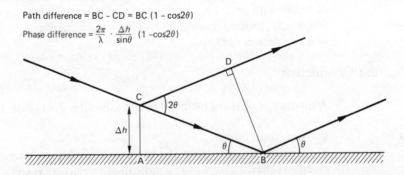

Path difference = BC - CD = BC $(1 - \cos 2\theta)$

Phase difference = $\dfrac{2\pi}{\lambda} \cdot \dfrac{\Delta h}{\sin\theta}$ $(1 - \cos 2\theta)$

Fig. 3.7   *Phase difference between two waves reflected from points differing in height by Δh*

Consider two waves reflected at points separated in height by Δh. Their phases are shifted with respect to each other (see Figure 3.7) by

$$\Delta\Phi = (4\pi\Delta h \sin\theta)/\lambda \tag{3.12}$$

where θ is the grazing angle of the ray path with the ground. If the surface is to behave like a *smooth surface*, then waves reflected at all points that participate in the reflection, i.e. the points within the first Fresnel zone, must be only very slightly shifted in phase with respect to each other. Rayleigh proposed to consider a surface as 'smooth' if

$$\Delta\Phi < \pi/2 \tag{3.13}$$

Using equation (3.12) this establishes the *Rayleigh criterion* which categorises the surface as smooth if

$$\Delta h < \lambda/(8 \sin \theta) \tag{3.14}$$

For radio waves in the HF band or for lower frequencies equation (3.14) shows that ground nearly always behaves like a perfectly smooth surface particularly with antennas near the ground. For VHF and higher frequencies, however, the effects of the ground irregularities start to become apparent. The greater the irregularities, the smaller the reflection coefficient.

The field scattered by a *rough surface* may be considered as the sum of two components: the specular component and the diffuse component. *Specular reflection* is a reflection of the same type as caused by a smooth surface; it is directional and obeys the laws of classical optics. Its phase is coherent and is the result of radiation of the points on the Fresnel ellipse. Its fluctuations have a relatively small amplitude. *Diffuse scattering* has little directivity and consequently takes place over a much larger area of the surface than the first Fresnel zone. Its phase is incoherent, and its fluctuations, which have a large amplitude, are Rayleigh distributed.

The definition of terrain 'roughness' is related to the validity of the assumption of either specular reflection or diffuse scattering. The magnitude of the experimental reflection coefficient[10] becomes less than the theoretical value as the grazing angle θ increases. When the roughness of a surface is less than $\lambda/(8 \sin \theta)$, specular reflection will occur and the theoretical values of reflection coefficient R should be valid. When the roughness is greater than $\lambda/(8 \sin \theta)$ diffuse reflection will occur and the theoretical values of R will not apply.

Reflection from a surface does not, of course, change abruptly from specular to diffuse as either λ, θ or Δh is varied. The change is a gradual one and the criterion of (3.14) is merely a good 'rule of thumb'.

3.4.4   Shadowing

Theoretical models for scattering from rough surfaces frequently neglect shadowing. In cases where (see Figure 3.8) significant proportions of the reflecting surface are not visible to both the receiving and transmitting antennas, R should be reduced to allow for shadowing. The *shadow factor*[11] represents the fraction of the first Fresnel zone area which is illuminated. Because of the complexity associated with the determination of the shadow factor and the frequent lack of sufficient statistical information concerning the reflection surface, the factor is often assumed to be unity.

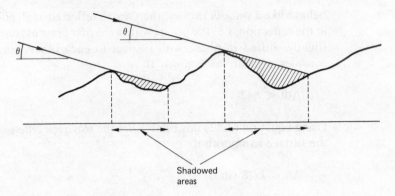

Fig. 3.8   *Shadowing effect caused by a rough surface*

3.4.5   Mountainous Terrain

Some paths can be treated as passing over a succession of crests, which may be replaced by knife-edges or cylinders, according to the sharpness of the crests.

On long paths, tropospheric-scatter may occur well above a mountain ridge, and the scattered and diffracted waves must be combined. With transmitting and receiving antennas elevated above the surrounding terrain, waves may be reflected both before and after diffraction. When a wave passes close to the ground, an additional transmission loss, caused by finite ground conductivity, may arise[12].

Mountain ridges can effectively reduce both transmission loss and the fading below the values to be expected in the absence of the obstacle. This occurs when the direct path is non-optical, but both transmitter and receiver can be seen from the top of the mountain. The phenomenon is known as *obstacle gain*. Measurements to compare propagation over two neighbouring paths of similar ranges and antenna heights, one of which has a mountain ridge which causes knife-edge diffraction and the other which is clear of obstacles, have confirmed that such gains do occur[13].

The direction of arrival of the strongest signal is not necessarily the direction of the great circle path between the transmitter and the receiver. This is most noticeable when the receiving station is very near to the diffraction ridge (a few kilometres). Thus in estimating the quality of transmission across a mountain ridge, consideration must be given not only to the profile of the terrain in the great circle path but also to the diffraction or scattering properties of the ridge outside this plane. However, if the mountain obstacle is only a little removed from the great circle path, it no longer introduces an appreciable gain.

3.4.6   Vegetation

In the 30 to 2000 MHz range the average additional attenuation through wooded terrain appears[14] to be proportional to exp $(-\beta d/\lambda)$ where β is a characteristic of the vegetation, d is distance and λ the wavelength. Considerable variation could be expected in these values as a result of the density of vegetation, the moisture content of leaves and the presence of snow on the branches.

At lower frequencies, those of interest here, the vegetation can be modelled as a weak, lossy dielectric slab. In this model[15] energy is coupled into a lateral wave travelling along the top of the slab, spreading with horizontal distance d according to a field strength which is proportional to $1/d^2$. Although the lateral wave continually leaks energy back into the vegetation, its strength can greatly exceed the wave produced along the direct propagation path which is subject to $\exp(-\beta d/\lambda)$ additional attenuation. The net effect is that the presence of vegetation produces a constant loss which seems to be independent of the distance between communications terminals.

3.5   Field Strength Computation

3.5.1   The Prediction Problem

Ever since the turn of the century when radio waves were first used for communication purposes, engineers have endeavoured to predict radio propagation characteristics for specific paths. Many mathematical techniques were developed to analyse the problem and simplified curves constructed for propagation over an ideal flat Earth. As theories were advanced and techniques became more sophisticated, calculations were made for a homogeneous curved Earth taking into account the tropospheric effects of refraction. It was appreciated that a real propagation path might differ considerably from the ideal situation and further studies enabled some specific deviations to be analysed. The problem of propagation over irregular terrain has been addressed by Ott[16]. Computer programs have been developed to compute quantitative effects for specific terrain profiles.

In the older ground wave propagation curves no account was taken of tropospheric refraction. The CCIR atlas[17] of ground wave propagation curves for frequencies above 30 MHz assumes that the tropospheric refractive index decreases linearly with height. The extended Sommerfeld theory[18] is restricted to such a linear variation.

In fact, the troposphere has a mean refractive index which varies exponentially with height thus[19]

$$n(h) = 1 + N \exp(-h/h_s) \tag{3.15}$$

where N is the difference of refractive index from unity at the surface and h_s is the scale height of the troposphere. This model is now accepted as the Reference Atmosphere of the CCIR[20] with recommended values of

$$\left. \begin{array}{l} N = 315 \times 10^{-6} \\ h_s = 7.35 \text{ km} \end{array} \right\} \tag{3.16}$$

In recent years a method of analysis has been developed[21] which enables the effects of a troposphere with an exponential height variation of refractive index to be taken into account. This method has been embodied in the computer prediction program GRWAVE which has been adopted by CCIR[22] to compute standard curves for ground wave propagation between 10 kHz and 30 MHz.

3.5.2   Field Strength Values

For horizontally polarised waves, ground losses are very much greater than for vertical polarisation. Thus signal strength decreases much more rapidly with distance than for vertical polarisation. Increasing the antenna height reduces the ground losses, thus reducing the difference between the two polarisations.

Because antennas at or near the surface are primarily of interest here, the effects of the horizontally polarised component can be ignored. Consequently estimates of field strength can be made using only the vertical component.

Based upon the simple model of a smooth, homogeneous Earth bounded by a troposphere with exponential height variation, calculations using the GRWAVE program are presented in Figure 3.9 for three types of ground. The ground parameters chosen are those recommended by the CCIR[23]:

a) Sea water $\sigma = 5$ S/m, $\epsilon = 80$
b) Wet ground $\sigma = 0.01$ S/m, $\epsilon = 30$
c) Medium dry ground $\sigma = 0.001$ S/m, $\epsilon = 15$

The reference radiator chosen is that recommended by the CCIR[1]. This is for a Hertzian vertical electric dipole with a dipole moment $5\,\lambda/2\pi$ ampere metres, giving a field of 0.3 V/m at a distance of 1 km on the surface of a perfectly conducting plane. (A vertical antenna shorter than one quarter wavelength is nearly equivalent to an ideal Hertzian vertical electric dipole.) This dipole moment is chosen so that the dipole would radiate 1 kW if the Earth were a perfectly conducting infinite plane. An example of the computed field strengths for a number of frequencies are shown in Figure 3.9. A more complete set of ground wave reference curves can be found elsewhere[21,22].

3.5.3   Variability of Ground Conditions

The effective values of the constants of the ground are determined, not only by the nature of the soil, but also by its moisture content and temperature, by the wave frequency, by the general geological structure of the ground and by the effective depth of penetration and lateral spread of the waves. The absorption of energy by vegetation, buildings and other objects on the surface must also be taken into consideration.

Values of dielectric constant and conductivity for sea water are relatively constant for different geographical areas. However, these parameters may vary considerably over different soil types so that values chosen in (b) and (c) of section 3.5.2 should be considered only as a representative sample.

In order to assess the effect of changing both ϵ and σ, it is convenient to consider two more types of ground with parameter values:

d) $\sigma = 0.01$ S/m, $\epsilon = 4$
e) $\sigma = 0.001$ S/m, $\epsilon = 4$

and to compare the results with the (b) and (c) of section 3.5.2. The typical differences caused by changing either σ or ϵ are summarised in Figure 3.10 and represent the spread of ground conditions of interest here. Curve A shows the

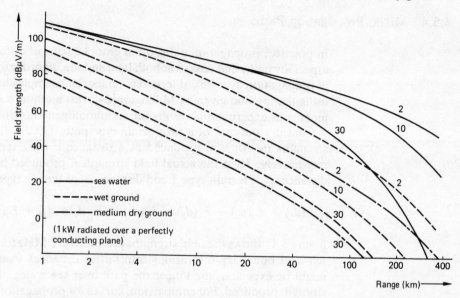

Fig. 3.9 *Ground wave propagation curves for different ground types (the numbers on the curves refer to frequency in MHz)*

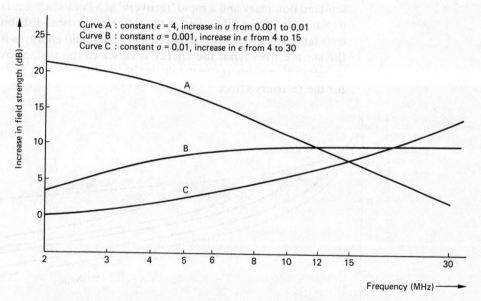

Fig. 3.10 *Effect of ground constants upon field strength*

effect on field strength of increasing conductivity (condition (*e*) to condition (*d*)). Note that the greater increase of field strength occurs at the lower frequencies. Curve B shows the increase caused when changing from condition (*e*) to condition (*c*), and curve C the increase caused when changing from condition (*d*) to condition (*b*). Both B and C represent increases in ϵ for constant σ. The effects are more marked at higher frequencies.

3.5.4   Mixed Propagation Paths

In practice, propagation paths often cross terrains of differing conductivity – especially paths that pass over both land and sea. In order to quantify the effect of propagation over mixed terrain a number of examples have been computed using the method given by Millington[24]; this method is in very good agreement with experimental results for an inhomogeneous smooth Earth.

Consider the case of a two-terrain type path. Let $E_1(d)$, $E_2(d)$ be the field strengths in dBµV/m produced at a distance d over terrain types 1 and 2 respectively. Then the actual field strength E produced by propagation over distance d_1 of terrain type 1 and distance d_2 of terrain type 2 is

$$E = \tfrac{1}{2}[E_1(d_1) + E_2(d_2) - E_1(d_2) - E_2(d_1) + E_1(d_1 + d_2) + E_2(d_1 + d_2)] \qquad (3.17)$$

Figure 3.11 shows the field strengths produced at 2 MHz for propagation over a sea-land boundary for various transmitter distances from the coastline. As might be expected, the longer the path over sea water, the greater the field strength produced. For comparison, curves for propagation over an all-sea and an all-land path are given.

More complicated situations are shown in Figures 3.12 and 3.13. Note particularly for the higher frequencies, the rapid decrease in field strength at a sea-land boundary and a rapid 'recovery' at a land-sea boundary. The physical explanation of the effect lies in the different vertical distribution of the field over land compared with that over sea. More field energy is found high above the surface over land; the energy is closer to the surface over sea. It is the redistribution of energy vertically in the vicinity of the boundary that accounts for the recovery effect.

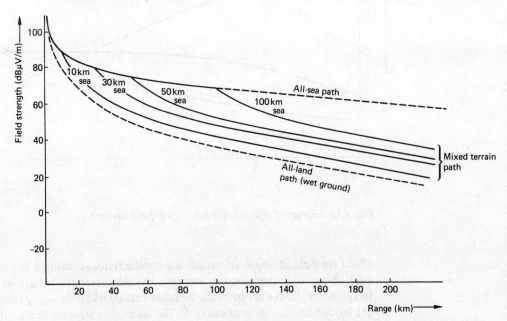

Fig. 3.11   *Field strength curves for 2 MHz vertically polarised wave over a sea-land path*

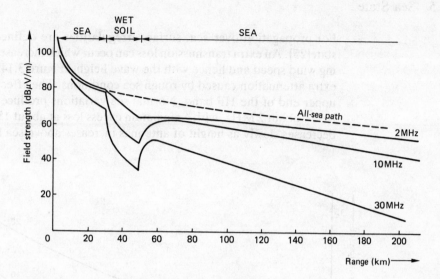

Fig. 3.12 *Field strength curves for a sea-land-sea path*

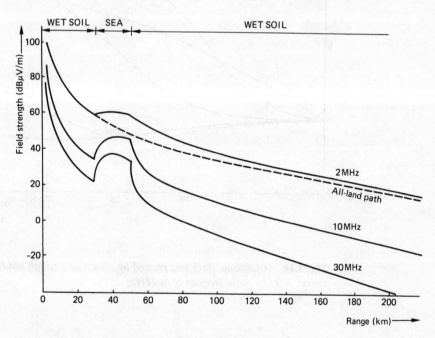

Fig. 3.13 *Field strength curves for a land-sea-land path*

At a considerable distance from the second boundary in Figures 3.12 and 3.13 (say greater than 100 km) the field strength has virtually returned to the value characteristic of an all-sea (Figure 3.12) or an all-land (Figure 3.13) path. In the vicinity of the boundaries, rapid variations in field strength may be experienced over short distances.

3.5.5 Sea State

For propagation over sea, surface irregularities are defined in terms of sea state[25]. An extra transmission loss can occur which increases with the prevailing wind speed and hence with the wave height. Figure 3.14 shows the typical extra attenuation caused by rough sea conditions. The effect is greatest at the upper end of the HF band. Normal sea variations produce negligible effects below about 2 MHz, with a maximum excess loss at about 15 MHz. The effect decreases slowly as height of antennas increases above sea level.

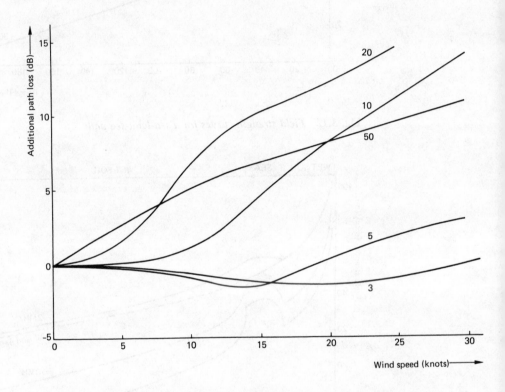

Fig. 3.14 *Additional path loss caused by effects of a rough sea (the numbers on the curves refer to radio frequency in MHz)*

3.5.6 Obstacles

For the HF band there is a simple method[26] of estimating the effect of large obstacles upon field strength. Figure 3.15 shows a nomogram used to estimate a *range reduction coefficient* of an HF signal in the presence of a large obstacle. At VHF, further complications arise including obstacle gain, spatial inhomogeneities in the diffraction field, and frequency and phase distortion of the signal. These effects are very small for frequencies in the HF band.

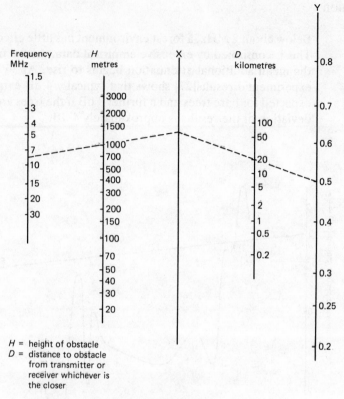

H = height of obstacle
D = distance to obstacle
from transmitter or
receiver whichever is
the closer

1. Join the frequency to height H and extend the line to X.
2. Join the point of intersection with X to distance D.
3. Extend the line to intersect Y and read reduction factor.

Fig. 3.15 *Nomogram to determine range reduction factor caused by obstacles*

The nomogram method of Figure 3.15 can give only an indication of the
order of effects produced, once the field has stabilised. To illustrate the rapid
nature of the field variation, even for lower frequencies, Figure 3.16 shows the
predicted field strength for a 1 MHz wave traversing terrain containing two
isolated mountains each 1000 m high. These calculations were performed using
the method of Ott[16]. Note how the field rapidly increases as the point of
reception moves up the mountain and then suddenly decreases as the point of
reception disappears over the top and down the other side. As distance
increases from the obstacle there is a slight recovery effect and gradually the
field stabilises. It is this stable value which Figure 3.15 aims to estimate.

In practice, a given propagation path may involve both mixed terrain con-
ductivities and obstacles. Figure 3.17 shows an example of a sea-land-sea path
with a hilly island. Note that the field strength may fall by almost 40 dB within
the space of a kilometre.

3.5.7 Vegetation

Below about 2 MHz, a forest environment has little effect on the ground wave. This is confirmed by extensive empirical data. As the frequency is increased, the mean additional attenuation begins to rise. At 30 MHz extrapolation of experimental results[27] shows that typically 4 dB extra attenuation may be expected for bare trees and a further 3 dB if the trees are in leaf. The standard deviation of the results is approximately 4 dB.

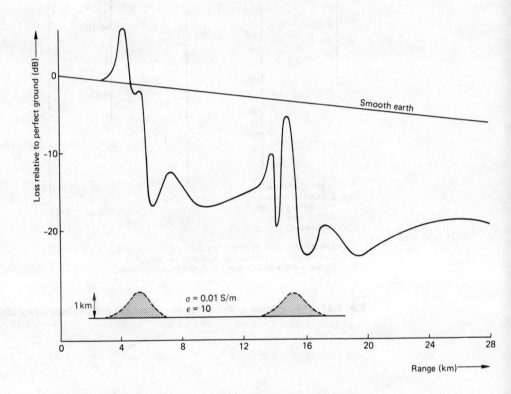

Fig. 3.16 *Propagation of a 1 MHz vertically polarised wave over two Gaussian-shaped ridges*

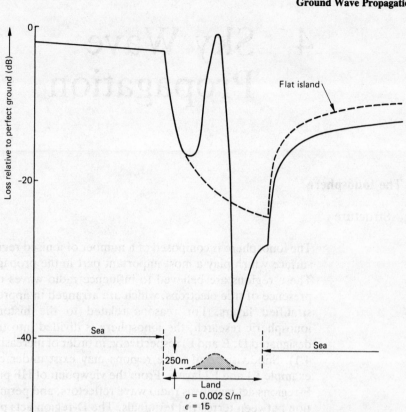

Fig. 3.17 *Propagation of a 10 MHz vertically polarised wave over a sea-land-sea path with hilly island*

4 Sky Wave Propagation

4.1 The Ionosphere

4.1.1 Structure

The ionosphere is composed of a number of ionised regions above the Earth's surface which play a most important part in the propagation of radio waves. These regions are believed to influence radio waves mainly because of the presence of free electrons, which are arranged in approximately horizontally stratified layers. For reasons related to the historical development of ionospheric research, the ionosphere is divided into three regions or layers designated D, E and F, respectively, in order of increasing altitude (see Figure 4.1). Subdivisions of these regions may exist under certain conditions, for example F1 and F2 layers. From the viewpoint of HF propagation, the E- and F-regions act mainly as radio wave reflectors, and permit long range propagation between terrestrial terminals. The D-region acts principally as an absorber, causing signal attenuation in the HF range, although VLF and ELF waves are reflected at D-region altitudes.

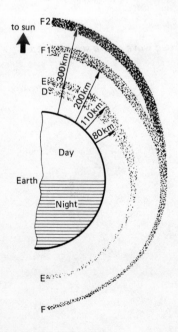

Fig. 4.1 *The ionospheric regions*

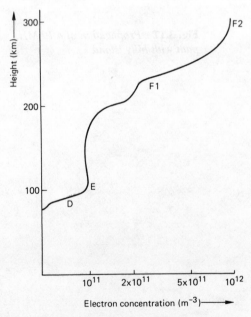

Fig. 4.2 *Typical daytime electron concentration distribution within the ionosphere*

The ionospheric structure varies widely over the Earth's surface[1], since the strength of the sun's radiation varies considerably with geographic latitude. An illustration of a daytime electron concentration height distribution in mid-latitudes is shown in Figure 4.2. Actual profiles vary over wide ranges and depend markedly upon time of day, season, sunspot number and whether or not the ionosphere is disturbed.

4.1.2 Ionisation

The principal source of ionisation in the ionosphere is electromagnetic radiation from the sun extending over the ultra-violet and X-ray portions of the spectrum. Other sources of ionisation are important, however, such as energetic charged particles of solar origin and galactic cosmic rays. The ionisation rate at various altitudes depends upon the intensity of the solar radiation (as a function of wavelength) and the ionisation efficiency of the neutral atmospheric gases. Since the sun's radiation is progressively absorbed in passing through the atmosphere, its residual ionising ability depends upon the length of the atmospheric path, and consequently upon the *solar zenith angle* (χ). The maximum ionisation rate occurs when the sun is overhead ($\chi = 0$), but geographic, diurnal and seasonal variations in the ionisation density are found. The production of free ionisation by solar radiation (and charged particles) is counter-balanced by ionisation loss processes, principally the collisional recombination of electrons and positive ions, and the attachment of electrons to neutral gas atoms and molecules.

4.1.3 D-region

The D-region spans the approximate altitude range 50–90 km with electron concentration increasing rapidly with altitude. The D-region electron density exhibits large diurnal variations. It has a maximum value shortly after local solar noon and a very small value at night. This diurnal variation is greatest in the altitude interval 70–90 km, with typical noon values of 10^8–10^9 electrons/m^3. There is a pronounced seasonal variation in D-region electron densities with a maximum in summer.

The relatively high density of the neutral atmosphere in the D-region causes the electron collision frequency to be correspondingly high ($\sim 2 \times 10^6$ sec^{-1} at 75 km). It is therefore in this region that the main absorption of energy from a propagating radio wave takes place.

4.1.4 E-region

The altitude range from 90–130 km constitutes the E-region and encompasses the so-called 'normal' and 'sporadic' E layers. The former is a regular layer which displays a strong solar zenith angle dependence with maximum density near noon and a seasonal maximum in summer. The altitude of maximum density is about 110 km, with a value of the order of 10^{11} electrons/m^3. At night only a small residual level of ionisation remains in the E-region. The solar cycle dependence exhibits a maximum layer density at solar sunspot maximum. The normal E layer is important for daytime HF propagation at distances less than 2000 km.

Sporadic E manifests itself in enhanced ionisation at E-region heights causing much greater critical frequencies when it is present. Its occurrence is strongly latitude dependent; in central European latitudes it is more frequent in summer than in winter and more frequent by day than by night. In high latitudes it is essentially a night-time phenomenon, in low latitudes a daytime one. Sporadic E occasionally prevents frequencies that normally penetrate the E layer from reaching higher layers and sometimes causes long-distance transmission at very high frequencies.

4.1.5   F-region

The F-region extends upwards from about 130 km and is divided into the F1 and F2 layers, although this distinction is only apparent during daytime. The F1 layer is the region between 130–210 km altitude, in which the maximum electron density is about $2 \times 10^{11}/m^3$. It exists only during daylight. This layer is occasionally the reflecting region for HF transmission, but more usually obliquely-incident waves that penetrate the E-region also penetrate the F1 layer and are reflected by the F2 layer. The F1 layer introduces additional absorption of such waves.

The F2 layer is the highest ionospheric layer, and usually exhibits the greatest electron density, which may range typically from $10^{12}/m^3$ in daytime to about $5 \times 10^{10}/m^3$ at night. The F2 layer is not well represented by a simple model since it is strongly influenced by winds, diffusion and other dynamic effects.

Ranging in height between about 250 to 400 km, the F2 layer is the principal reflecting region for long distance HF communication. Height and ionisation density vary diurnally, seasonally and over the sunspot cycle. Ionisation does not follow the solar zenith angle in any fashion since with such low molecular collision rates the medium can store received solar energy for many hours. At night the F1 layer merges with the F2 layer at a height of about 300 km. The absence of the F1 layer and reduction in absorption of the E-region causes night-time field intensities and noise to be generally higher than during daylight.

Some special features of the F-region occur at low and high latitudes; these can have important effects upon radio wave propagation. Near the equator significant latitudinal gradients exist in the F-region ionisation whilst at high latitudes there is a region of strongly depressed electron density.

4.1.6   Ionospheric Disturbances

The term *ionospheric disturbance* is used to cover a wide variety of conditions that show some departure from the usual state[2]. As far as radio communications are concerned the most important disturbances are those which are associated in some way or other with a solar flare (see Figure 4.3), namely

1   Sudden ionospheric disturbances (SID)
2   Ionospheric storms
3   Polar cap absorption events (PCA).

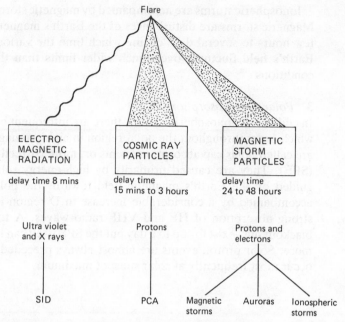

Fig. 4.3 *Some terrestrial effects of a solar flare*

The reason these disturbances are important is that they often result in interruption of communications for periods lasting from a few minutes to several days. This may occur because of enhanced D-region absorption or because of depression of F2 layer electron densities, both of which may result in a loss of signal. It is therefore important to understand the physical causes of such disturbances.

1 Sudden Ionospheric Disturbances (SID)
Occasionally a sudden outburst on the sun, a solar X-ray (SXR) flare event (often known simply as a solar flare), produces abnormally high ionisation densities in the D-region.

This ionospheric phenomenon, called a sudden ionospheric disturbance, results in a sudden increase in the absorption of MF, HF and VHF radio waves, although the field strength of LF and VLF waves can often be enhanced. Because SIDs are produced by direct X-ray and ultra violet radiations from the sun, they occur only on the sunlit side of the Earth, and are most frequent at solar maximum.

2 Ionospheric Storms
Ionospheric storms, which may last for several days, are caused by streams of charged particles of solar origin deflected by the Earth's magnetic field towards the auroral zones. They are accompanied by an increase in electron density in the D-region and an expansion and diffusion of the F2 layer which can result in decreased critical frequencies and increased layer heights. The effects of ionospheric storms are most severe at solar maximum but are more significant relative to communications at solar minimum.

Ionospheric storms are accompanied by magnetic storms and auroral effects. Magnetic storms are disturbances of the Earth's magnetic field lasting from a few hours to several days during which time the various components of the Earth's field fluctuate over much wider limits than they do under normal conditions.

3 *Polar Cap Absorption* (PCA)

In addition to ionospheric storms, there are infrequent but major disturbances which occur throughout the polar region in high geomagnetic latitudes. These are called polar cap absorption events or, more correctly, solar proton events (SPE). They are caused primarily by high-energy solar protons which are guided by the Earth's magnetic field towards the polar regions. SPEs are accompanied by a considerable increase in D-region ionisation resulting in strong absorption of HF and VHF radio waves. A total polar radio wave blackout may exist for up to a day but the SPE itself can last for up to a week or more. Solar proton events are almost always preceded by a major flare and occur most frequently at solar sunspot maximum.

4.2   Wave Propagation in the Ionosphere

4.2.1   Physical Processes

When a radio wave travels through the ionosphere its electric field imparts an oscillatory motion to the electrons, which re-radiate like miniature antennas. The re-radiation modifies the velocity of propagation of the wave[3]. Refraction occurs when the electron concentration changes, so that waves are refracted back towards the Earth if their frequency is not too high. If the Earth's magnetic field were absent the oscillatory motion of the electrons would be parallel to the direction of the electric field of the incident wave and the re-radiated waves would have the same polarisation as the incident wave; the wave would propagate through the ionosphere without change of polarisation. The Earth's magnetic field modifies the oscillatory motion of the electrons causing them to move in complicated orbits. Their re-radiation is not in general of the same polarisation as the incident wave. Thus the resultant wave polarisation changes continuously as the wave traverses the ionosphere.

In its passage through the ionosphere a radio wave imparts a minute, but significant, quantity of energy to each electron. The constituents of the ionosphere are in continuous random motion and consequently collisions between electrons and heavier particles occur frequently. When such a collision takes place the ordered energy supplied to an electron by the radio wave is converted into disordered (heat) energy and the wave is therefore attenuated. The more frequently the electrons collide the greater is the attenuation.

4.2.2   Reflection at Oblique Incidence

If the effects of the Earth's magnetic field are ignored then the refractive index *n* of the ionosphere is given by[2]

$$n^2 = 1 - (f_N/f)^2 \tag{4.1}$$

where f is the wave frequency and f_N is the plasma frequency which is proportional to the square root of the electron concentration. As the refractive index in vacuo is unity, a wave ascending into the ionosphere encounters a region where the refractive index gradually falls as the electron concentration increases with height. If the ionised layer is sufficiently thick refraction will continue until the angle of refraction reaches 90°. The ray will then have reached its highest point and will start its downward journey back to Earth. In practice, of course, the variation of electron concentration is continuous and the path of the ray will likewise be a continuous curve.

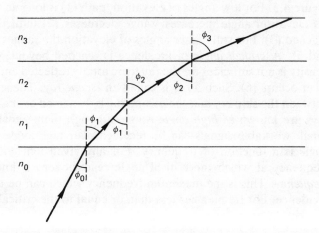

Fig. 4.4 *Refraction in a layered medium*

If, for simplicity, the layer is considered to be divided into a number of thin strips of constant electron density, with each strip having a greater electron density than the one beneath it, then successive refraction at the boundaries between the strips will cause bending of the ray in the manner shown in Figure 4.4. Applying Snell's Law at the boundary of each strip gives

$$n_0 \sin \phi_0 = n_1 \sin \phi_1 = \ldots = n_n \sin \phi_n \tag{4.2}$$

Thus a ray entering the ionosphere at an angle of incidence ϕ_0 will be reflected at a height where the ionisation is such that n has the value

$$n = \sin \phi_0 \tag{4.3}$$

At vertical incidence the reflection condition occurs when n equals zero and from equation (4.1) this occurs where $f = f_N$. If $f = f_v$ represents the vertically incident frequency reflected at the level where the plasma frequency is f_N then for the obliquely incident wave

$$\sin^2 \phi_0 = 1 - (f_N/f)^2 = 1 - (f_v/f)^2 \tag{4.4}$$

and therefore

$$f = f_v \sec \phi_0 \tag{4.5}$$

Thus a frequency f incident on the ionosphere at an angle ϕ_0 will be reflected from the same electron density (true height) as the equivalent vertical incidence frequency $f_v = f \cos \phi_0$; hence a given ionospheric layer will always reflect higher frequencies at oblique incidence than at vertical incidence.

4.2.3   Ray Paths

For a fixed frequency the paths of rays leaving the transmitter are shown in Figure 4.5. For low angles of elevation, path (1) is long and the range is large. As elevation angle increases, range decreases (2) until the skip distance is reached (3). For still higher angles of elevation the ranges increase rapidly (4 and 5). A *critical angle* of incidence is reached beyond which the electron density is not sufficient for the ray to be totally reflected and penetration of the layer occurs (6). Such a ray is called an *escape ray*. The small bundle of rays between the skip ray and the escape ray is dispersed over a great range. These rays are known as *high angle rays*. Although their signal strengths may be small, workable signals can be received over high angle paths. The critical angle is a function of frequency. For any given ionisation distribution the frequency at which the critical angle reaches zero is known as the *critical frequency*. This is the maximum frequency which can be reflected at vertical incidence. For frequencies less than or equal to the critical frequency there is no skip distance.

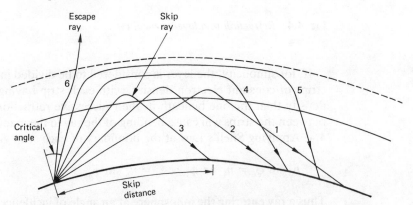

Fig. 4.5   *Ray paths as a function of elevation angle for a fixed frequency*

4.2.4   Virtual Height

Ionospheric characteristics are determined from measurements of the critical frequencies of the various layers. The most common method is that in which a transmitter radiates vertically upwards in short pulses. A nearby receiver picks up both the direct signal and that reflected from the ionosphere and measures the time difference between them. From this measurement the height at which reflection is taking place is calculated. The height actually determined is the *virtual height h'*. This is the height from which the wave would appear to be reflected, if the ionosphere were to be replaced by a perfectly reflecting surface at such a level that would imply the wave velocity were equal to the velocity of light. As the group velocity of the wave in the ionosphere is always less than the velocity of light in vacuo, the virtual height is always greater than the actual height h. A theorem[2] due to Breit and Tuve states that the virtual (or equivalent) path P' between a transmitter T and a receiver R is given by the length of the equivalent triangle TAR (see Figure 4.6).

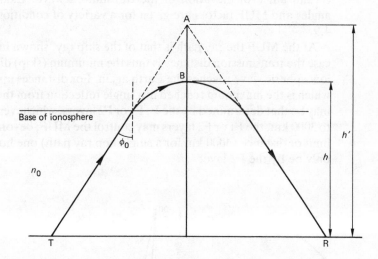

Fig. 4.6   *True path TBR and virtual (equivalent) path TAR*

If a pulse which is vertically incident on the ionosphere is assumed to travel with the free space velocity c, the equivalent path P' can be written

$$P' = 2h' = ct \tag{4.6}$$

where h' is the equivalent height of reflection (see Figure 4.6). A theorem due to Martyn[3] states that the virtual height of reflection of an obliquely incident wave is the same as that of an equivalent vertical wave:

$$P'(f) = 2h'(f_v) \sec \phi_0 \tag{4.7}$$

or $\cos \phi_0 P'(f) = 2h'(f \cos \phi_0)$

For a curved Earth and curved ionosphere the equivalence theorems are no longer valid. Sec ϕ_0 depends not only on h' and range but also on the electron density distribution. It will therefore change as the ionosphere changes, for example with time of day. The expressions to determine sec ϕ_0 become complicated and for most purposes it is sufficiently accurate to introduce a correction factor k so that the secant law in equation (4.5) now becomes

$$f = k f_v \sec \phi_0 \qquad (4.8)$$

4.2.5   Maximum Usable Frequency

The maximum usable frequency (MUF) for any given transmission distance is calculated from the critical frequency by multiplying by an MUF factor which is a function of the transmission distance. This distance is associated with a certain angle of elevation of the transmitted wave. Examples of elevation angles and MUF factors are given for a variety of conditions in Figures 4.7 to 4.9.

At the MUF the ray path is that of the skip ray, shown in Figure 4.5. In this case the transmission distance is thus the minimum (skip) distance at which the ionospheric wave reaches the Earth again. For distances up to about 2000 km, which is the maximum reached in a single reflection from the E layer, the MUF may be that determined by the E, F1 or F2 layers, whichever is the greatest. Up to 3000 km, the F1 or F2 layers may control the MUF; beyond this and up to the limit of distance (4000 km for a single hop ray path) one hop transmission can only be by the F2 layer.

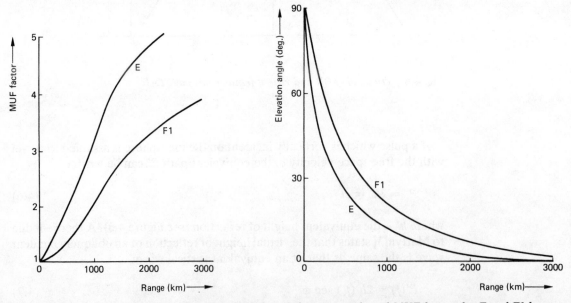

Fig. 4.7   *Range dependence of elevation angle and MUF factors for E and F1 layers*

In practice the MUF is not a sharp limit and propagation is often possible on frequencies greater than the classical MUF. The maximum operational frequency may be appreciably higher than the MUF. This extension arises since neither the ground nor the ionosphere are smooth reflectors, as assumed in the simplified theory. Scattering from irregularities will therefore allow signals to propagate to distances beyond the limit of the refracted wave. Ionospheric tilts can also play an important role in extending the operational frequency above the MUF.

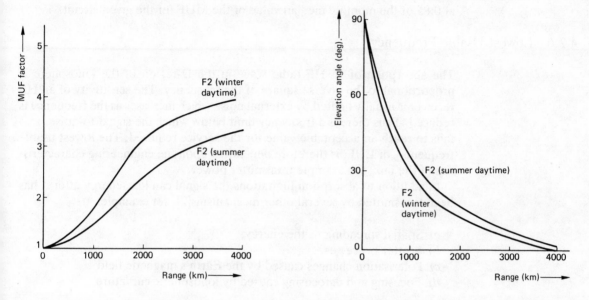

Fig. 4.8 *Range dependence of elevation angle and MUF factors for F2 layer (daytime)*

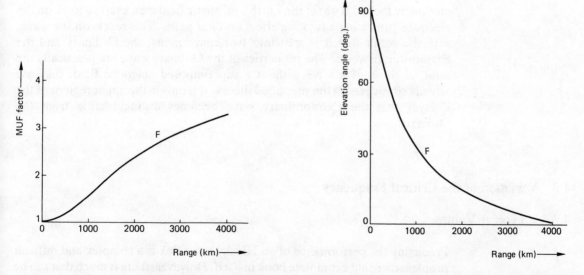

Fig. 4.9 *Range dependence of elevation angle and MUF factors for F layer (nighttime)*

The recommended upper limit, or optimum working frequency, is selected below the MUF to provide some margin for ionospheric irregularities and turbulence as well as for the statistical deviation of day-to-day ionospheric characteristics from the predicted monthly median value. The optimum working frequency, known as the FOT, from the French initials, is fixed empirically at 0.85 of the monthly median value of the MUF for the given circuit.

4.2.6 Lowest Usable Frequency

The absorption of an HF radio wave in the D-region of the ionosphere is proportional to the inverse square of the frequency. The sensitivity of an HF receiver is usually limited by external noise which increases as the frequency is reduced. Thus there is a frequency limit below which the signal-to-noise ratio fails to reach an acceptable value for the service required. The lowest usable frequency, or LUF, is therefore dependent upon the engineering characteristics of the link, for example transmitter power.

In addition to absorption limitations the signal can lose energy after it has been transmitted by several other mechanisms[2], for example:

a) Spatial spreading of the energy.
b) Scatter processes.
c) Polarisation changes caused by the Earth's magnetic field.
d) Focusing and defocusing caused by ionospheric curvature.

These processes complicate the evaluation of the expected performance of the circuit.

4.2.7 Effect of the Earth's Magnetic Field

As explained in Section 4.2.1, the electrons within the ionosphere are set in motion by the radio wave; the Earth's magnetic field then exerts a force on the electrons producing a twisting effect on their paths. This reacts on the wave, with the result that it is split into two components, the Ordinary and the Extraordinary waves. The properties of the Ordinary wave are practically the same as those of a wave without a superimposed magnetic field, the case already considered in the simplified theory. It is only in the upper regions of the F layer that the Extraordinary wave becomes distinguishable from the Ordinary.

4.3 Variation of the Critical Frequency

4.3.1 Typical Values

Predicting the performance of an HF sky wave link is a complex and difficult problem meriting a complete book in itself. However, there is much that can be learned from examining the dependencies of some of the relevant characteristics, of which the critical frequency is particularly important.

IONOSPHERIC LAYER	TIME	ELECTRON CONCENTRA- TION (m^{-3})	CRITICAL FREQUENCY (MHz)
D	Noon	$10^8 - 10^9$	0.09 – 0.28
E	Noon	10^{11}	2.8
F1	Noon	2×10^{11}	4.0
F2	Day	10^{12}	9.0
F2	Night	5×10^{10}	2.0

Table 4.1   Typical maximum electron concentrations and layer critical frequencies.

The critical frequencies of the E and F2 layers of the ionosphere, known respectively as f_oE and f_oF2, are the highest frequencies capable of being reflected from the two regions; they are related to the maximum electron densities in those regions. The value of f_oF2 is always greater than that of f_oE because the electron concentration in the F layer is considerably greater than that in the E layer; it is often the case that at night there is no significant E layer at all. The parameter f_oF2, then, is of primary concern and, for vertical incidence, is synonymous with the MUF.

To a good approximation, the plasma frequency f_N is given by

$$f_N = 9N^{1/2} \tag{4.9}$$

where N is the electron density in electrons per cubic metre and f_N is in hertz. If N_m is the maximum electron density in a given ionised layer then all waves whose frequency is less than the plasma frequency (for an electron concentration N_m) that enter the ionosphere at vertical incidence will be reflected back to Earth. The critical frequency is given approximately by

$$f_o = 9N_m^{1/2} \tag{4.10}$$

The critical frequencies of the D, E and F layers can be calculated from the corresponding values of N_m given in Section 4.1. The results are shown in Table 4.1.

4.3.2   Solar Cycle Dependence

An analysis of vertical soundings made by the Rutherford-Appleton Laboratory at Slough over an eleven year period shows the solar cycle dependency of the critical frequencies. The data is representative of the range of conditions to be experienced in mid-European latitudes. Values of smoothed (12 month running average) sunspot number are plotted in Figure 4.10. It can be seen that 1969 and 1980 coincide with high sunspot activity, 1976 with low sunspot activity. Figure 4.11 shows the monthly mean noon critical frequencies of the E and F2 layers as observed at Slough over the last solar cycle. Note in particular:

a) There is a marked correlation of F2 layer critical frequencies with the eleven year solar cycle, superimposed as the broken curve in Figure 4.11.
b) The seasonal variation of f_oE is in phase with solar zenith angle (ie greater in the summer months) whereas f_oF2 is in antiphase. This is known as the *winter anomaly*[2].

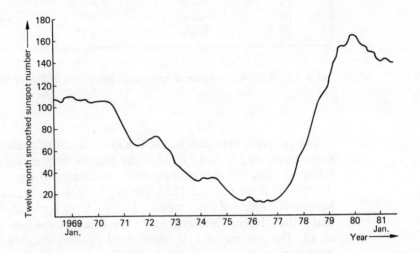

Fig. 4.10   *Twelve month smoothed sunspot numbers for the 1970s*

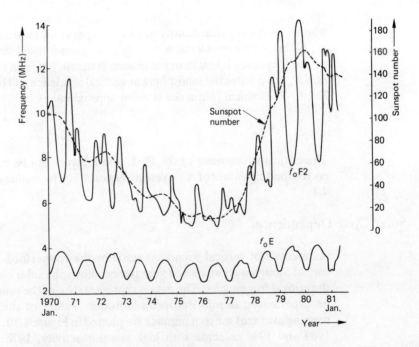

Fig. 4.11   *Monthly mean of the noon critical frequencies observed at Slough 1970-81 with superimposed sunspot activity*

The F2 layer is the most important layer from the viewpoint of HF radio communications; unfortunately it is also the most variable. Unlike the E layer (and the F1 layer, when present) the F2 layer does not follow a simple solar zenith angle law, either diurnally or seasonally. Thus although during January the sun at Slough is at a low elevation compared with July, Figure 4.11 shows that the January noon f_oF2 may be as much as twice the summer value. This implies a peak electron density ratio of more than four to one. This winter anomaly effect occurs in daytime only.

4.3.3   Annual, Seasonal and Diurnal Variations

To illustrate the behaviour of the critical frequency of the F2 layer during a solar cycle, three years 1970, 1973 and 1976 have been chosen as representative of high, medium and low sunspot activity. The monthly median F2 critical frequencies as measured at Slough for the three years of interest based upon hour of the day are plotted in Figure 4.12; for comparison the E layer critical frequencies f_oE are shown in Figure 4.13. The difference between high and low sunspot activity is quite significant. These variations are further highlighted in Figure 4.14 which plots the distributions for each year, and Figure 4.15 which shows the cumulative distributions. It is apparent that critical frequencies during low sunspot activity are limited to a maximum of 6 MHz, whereas for high sunspot activity critical frequencies higher than 6 MHz occur for up to 70% of the time.

Figure 4.16 shows the annual variation of monthly median f_oF2 for four months of the year. For daytime conditions, the greatest variation over the solar cycle occurs in winter, the least variation in the summer. By contrast, night-time conditions in the winter are largely unaffected by sunspot number.

Figure 4.17 shows seasonal variation of monthly median f_oF2 for three years in the last solar cycle. As might be expected the largest variation over the year occurs during periods of high sunspot activity (1970).

Figure 4.18 shows diurnal variation of monthly median f_oF2 for four months and three separate years. The trend from low to high sunspot number is clearly apparent. Diurnal variations are most marked during winter months.

From the above information it is possible to deduce the likely median critical frequency for a given set of conditions relevant to mid-latitude. Table 4.2 summarises these values, together with their expected day-to-day variations. Operations in higher latitudes will tend to incur lower critical frequencies and greater variabilities.

There are occasions (in the winter of low sunspot number years) when f_oF2 does not reach even 2 MHz. In the last sunspot cycle this occurred, for example, in January, February and March of 1976 between 0500 and 0700 GMT. The required operating frequency under such conditions may be outside the frequency of operation of the HF radio equipment. The values given in Table 4.2 take no account of sporadic E or ionospheric storm effects. On occasions, the highest critical frequency in the presence of sporadic E has reached 15 MHz, compared with approximately 10 MHz maximum critical frequency for the normal conditions when reflections are returned from the F2 layer.

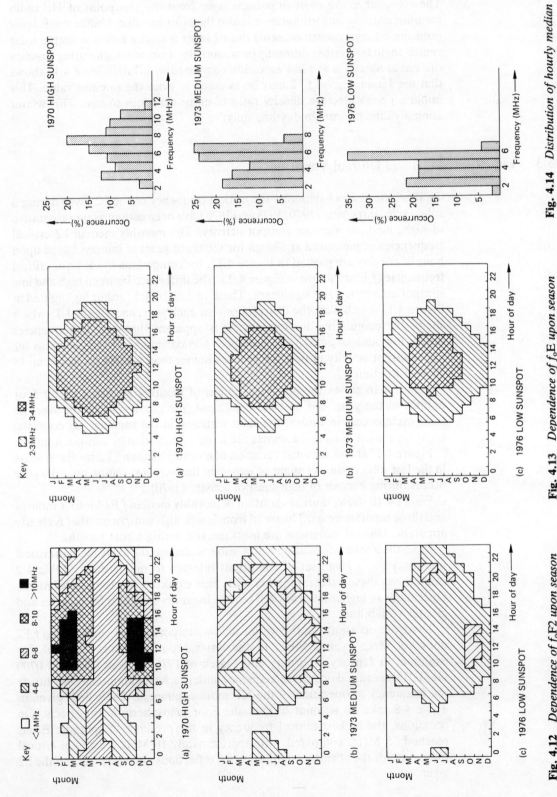

Fig. 4.12 *Dependence of f_oF2 upon season and time of day observed at Slough*

Fig. 4.13 *Dependence of f_oE upon season and time of day observed at Slough*

Fig. 4.14 *Distribution of hourly median f_oF2 values at Slough*

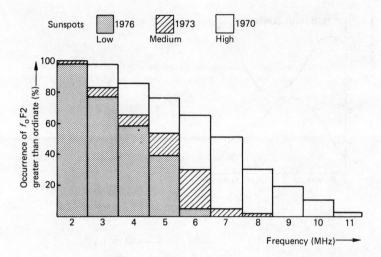

Fig. 4.15 *Cumulative distribution of hourly median f_oF2 at Slough*

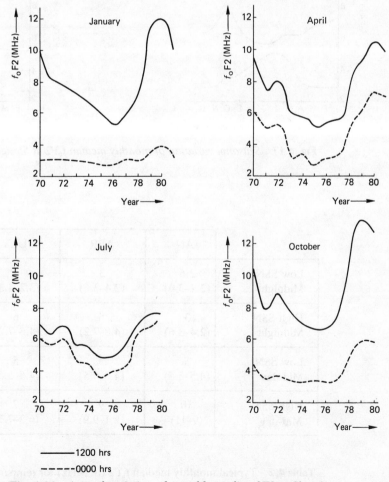

Fig. 4.16 *Annual variation of monthly median f_oF2 at Slough*

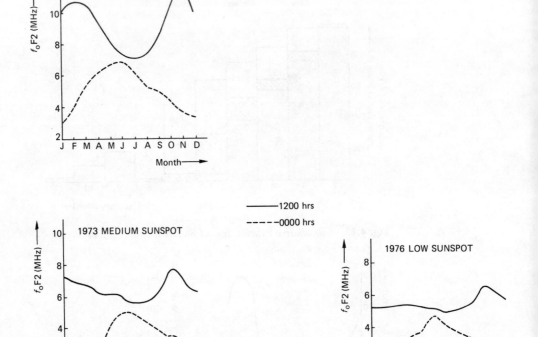

Fig. 4.17   *Seasonal variation of monthly median f_0F2 at Slough*

	JAN	APR	JULY	OCT
Low SSN Midnight	2.6 (2.2–3.0)	3 (2.4–3.6)	4 (3.2–4.8)	3 (2.4–3.6)
High SSN Midnight	3 (2.4–3.6)	6 (4.8–7.2)	6 (4.8–7.2)	4 (3.2–4.8)
Low SSN Mid-day	5 (4.5–5.5)	5 (4.5–5.5)	5 (4.5–5.5)	6 (4.4–5.6)
High SSN Mid-day	10 (9–11)	9 (8.1–9.9)	7 (6.3–7.7)	9 (8.1–9.9)

Table 4.2   Typical monthly median f_0F2 (in MHz) for temperate latitudes, with the probable range in day-to-day variations

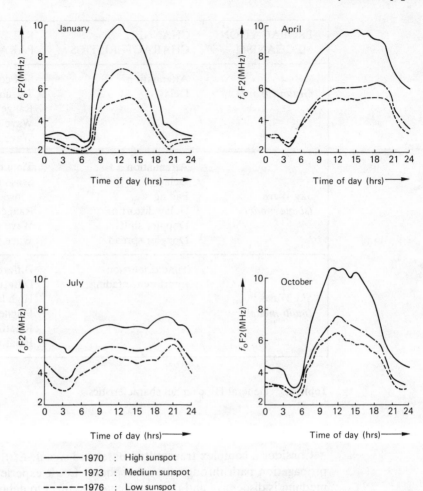

Fig. 4.18 *Diurnal variation of monthly median f_oF2 at Slough*

4.4 Characteristics of the Received Signal[4]

4.4.1 Components

Ground wave communication is more straightforward than sky wave; it can be assumed that the ground wave is merely an attenuated, delayed but otherwise undistorted version of the transmitted signal. Ionospheric sky wave returns, however, in addition to experiencing a much greater variability of attenuation and delay, also suffer from fading, frequency (Doppler) shifting and spreading, time dispersion and delay distortion. These features are summarised in Table 4.3 and are discussed in detail in this section.

PROPAGATION MECHANISM	CHANNEL CHARACTERISTICS	RELEVANT PARAMETERS
Ground Wave	Attenuation Delay	Soil conductivity Terrain type Range Wave polarisation Wave frequency
Sky Wave (single mode)	Attenuation Delay Fading Delay distortion Doppler shift Doppler spread	Time of day Season Sunspot activity Range Wave polarisation Wave frequency
Sky Wave (multi-mode)	Time dispersion Interference fading	Different hops Different modes High/low angle rays Magneto-ionic effects Relative attenuation Relative delay

Table 4.3   General HF channel characteristics

Consider a complex transmitted baseband signal, $E(t)$, traversing a single propagation path through the ionosphere. Let it experience a delay τ. The medium is dispersive and thus the signal is subject to delay distortion, caused by the fact that the delay is a function of frequency. This distorted waveform is denoted by $E'(t)$. In addition, the signal experiences attenuation and random fading. This can be represented by multiplying the delayed, distorted signal by a random gain $G(A,v,\sigma,t)$ where A characterises the attenuation ($0 \leqslant |A| \leqslant 1$) and v,σ represent the fading in terms of a frequency shift and spread respectively. The received signal $E_R(t)$ thus becomes

$$E_R(t) = G(A,v,\sigma,t)\,E'(t - \tau) \tag{4.11}$$

Components of this signal may be returned from both the E-region and F-region of the ionosphere (the latter may include both high and low angle ray paths). There are sky wave returns for the Ordinary and Extraordinary magneto-ionic components and for multiple hop paths (Figure 4.19). Although many propagation modes are possible all but a few experience a large attenuation; the number of 'effective' modes is generally small.

Each mode has different values of the characteristics of equation (4.11). For the jth mode, the received signal is

$$E_{Rj}(t) = G_j(A_j,v_j,\sigma_j,t)\,E'_j\,(t - \tau_j) \tag{4.12}$$

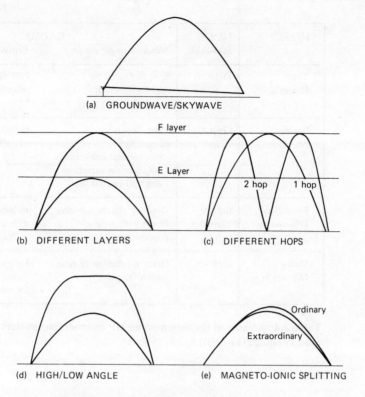

(a) GROUNDWAVE/SKYWAVE

F layer

E Layer

2 hop   1 hop

(b) DIFFERENT LAYERS   (c) DIFFERENT HOPS

Ordinary

Extraordinary

(d) HIGH/LOW ANGLE   (e) MAGNETO-IONIC SPLITTING

Fig. 4.19 *Causes of multipath propagation*

Consider now the ground wave. This can be assumed to experience a delay τ_g, and a non-random gain, but no distortion. Thus

$E'(t - \tau)$ becomes $E(t - \tau_g)$

G becomes A_g $(0 \leqslant |A_g| \leqslant 1)$

The total received signal is then

$$E_R(t) = A_g E(t - \tau_g) + \sum_{j=1}^{M} G_j(A_j, v_j, \sigma_j, t) E'_j(t - \tau_j) \qquad (4.13)$$

where M represents the number of effective sky wave modes. The relevant phenomena are summarised in Table 4.4. Each is now discussed in more detail in terms of its cause, magnitude, variability and resultant effect.

EFFECT	MODE PARAM.	CAUSE	
		Within a single mode	*Between different modes*
Time Dispersion	Delay τ_j	Spread caused by slightly different ray paths.	Different for each propagation mode.
Fading	Gain G_j	Time dependent caused by - ionospheric movements - polarisation variations - absorption changes	Different time dependence of each mode.
Frequency Dispersion	Shift ν_j Spread σ_j	Doppler effects non-zero. Phase path is time-dependent.	Relative phase between different modes changes with time
Delay Distortion	Delay τ_j	Delay a function of time and/or frequency.	Delay may have different frequency/time dependence for each mode.

Table 4.4   Causes of distortion on an HF channel (parameters referenced are those used in equation (4.13))

4.4.2   Multipath Propagation and Time Dispersion

Multipath characteristics can be described by the dispersion produced in the unit impulse response of the medium. *Time dispersion* can result from one or more of the following (see Figure 4.19):

a) Ground wave and sky wave paths.
b) Sky wave returns from different ionospheric layers.
c) Sky wave returns involving different numbers of hops.
d) High and low angle sky wave paths.
e) Splitting of the magneto-ionic components, Ordinary and Extraordinary, resulting from the effects of the Earth's magnetic field.

Each propagation path or mode has its own characteristic *group delay*. The time dispersion of the medium is caused by the difference in group delays between the different modes; it can give rise to intersymbol interference when the signalling rate becomes comparable with the relative multipath delays. The maximum serial data transmission rate is thus limited to the reciprocal of the range of multipath propagation times. This is itself a function of frequency, path length, geographical location, local time, season and sunspot activity. The data rate can be maximised by working close to the MUF.

As the operating frequency is decreased from the MUF, a frequency is reached at which the spread is a maximum. For a 2500 km path, the maximum time dispersion has been shown[2] to be about 3 ms; for 1000 km it increases to 5 ms and for 200 km it is about 8 ms. However, under conditions of maximum time dispersion it is more likely that the relative strengths of the propagation modes are considerably different.

Under some conditions, the transmission rate could be increased by a factor of 100 over the normal values[2] by judicious choice of operating frequency. In practice however, the upper limit is approximately 200 bits per second when conventional detection equipment is used. Even within a single mode of propagation, there remains an approximately 100 μs spread due to the slightly different constituent ray trajectories caused by roughness of the ionospheric layers and non-zero antenna beamwidths. The time dispersion can be much greater under anomalous conditions, such as spread F, when the ionosphere contains many irregularities.

Multipath has several important effects upon a given communications technique and its associated equipment when transmitting high-speed digital HF data. These are discussed in detail in Chapter 9, but are summarised briefly as follows:

a) The equipment is more complex, with special modems, diversity combining, etc. For example, in phase shift keyed systems, abrupt phase changes occur as successive modes reach the receiver, necessitating the provision of a guard interval at the end of each signalling period. In band-limited systems using multitone signalling, this reduces the number of tones available, since greater frequency separation is required.

b) The channel performance in terms of error rate is degraded, as a result of intersymbol interference, and high error rates may occur even at high signal-to-noise ratios.

c) The choice of operating frequency is limited to a small frequency band below the MUF. Working at frequencies too far below the MUF increases the likelihood of encountering larger multipath delays.

4.4.3 Fading

Sky wave signals fluctuate characteristically in amplitude and phase. The amplitude of the signal at a fixed receiver would remain steady no matter how irregular the ionosphere if it were a static medium. The width of the received power spectrum (i.e. the fading rate) is thus related to changes in the ionosphere.

There are a number of different kinds of fading, defined according to their origin. The main causes are movements and changes of curvature of the ionospheric reflector, rotation of the axes of the received polarisation ellipse, time variations of absorption and changes in electron density. In addition to these effects which may be produced independently for each mode, more significant fading may be caused by interference between two or more modes, particularly when they are roughly of equal amplitude. The different types of fading, with their typical fading rates, are summarised in Table 4.5.

Figure 4.20 presents some average fading rates for a typical HF channel at mid-latitudes[5]. Measurements were taken on a point-to-point link over a 2000 km path. It is clear that, particularly for the dawn and evening periods, the 10–50 fades per minute grouping is by far the most common. This is caused by interference between different sky wave modes. For midday, the results are spread rather more evenly from 0 to 50 fades per minute, but again higher rates of fading are infrequent. Also shown in Figure 4.20 is fade depth; fades of less than 10 dB occur most frequently.

CAUSE	FADING TYPE	FADING PERIOD	REMARKS
Small scale irregularities in F-region	Flutter	10–100 ms	Associated with spread-F.
Movement of irregularities in ionosphere	Diffraction	10–20 s	Follows a Rayleigh distribution.
Rotation of axes of polarisation ellipse	Polarisation	10–100 s	Both magneto-ionic components must be present.
Time variation of the MUF	Skip	usually non-periodic	Avoided by working well below the MUF.
Curvature of reflecting layer	Focusing	15–30 min	
Time variation of ionospheric absorption	Absorption	60 min	Greatest effects at sunset and sunrise.
Comparable strengths of different modes of propagation	Groundwave/ sky wave	2–10 s	
	Sky waves	1–5 s	
	High/low angle rays	0.5–2 s	
	Magneto-ionic splitting	10–40 s	

Table 4.5   Summary of HF fading characteristics

For a two-path channel with relative delay d seconds, troughs in the amplitude-frequency response are separated by $1/d$ Hz and give rise to frequency selective fading; signals with bandwidths greater than $1/d$ Hz are thus required for in-band frequency diversity. The $1/d$ Hz bandwidth is known as the *correlation bandwidth* and is given for different types of fading in Table 4.5. As the distance between two closely spaced receivers is increased, the correlation coefficient between their respective received signals decreases. The distance at which the coefficient drops to $1/e$ is called the *correlation distance*; it is of the order of a few wavelengths for sky wave reception (i.e. greater than 100 m at HF) and indicates the minimum antenna spacing required for space diversity reception.

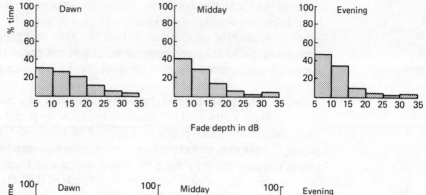

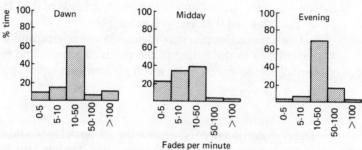

Fig. 4.20 *Typical fade rates and depths of an HF signal as received by a monopole*

4.4.4 Frequency Dispersion

For any given single propagation path, a shift v_j in frequency can be caused by time variation of

    a) height of the reflecting layer
    b) electron density (and hence refractive index) along the path.

Thus, if ψ is the phase angle of a ray path at time t, then

$$v_j = -\frac{f}{c}\frac{d\psi}{dt} \tag{4.14}$$

for a fixed transmitter and receiver.

The frequency (or Doppler) shifts experienced at night[6] are small compared to daytime effects, whilst relatively large positive values occur at sunrise and large negative values at sunset. On 'quiet' days, values range[7] from 0.01–1 Hz for single hop paths. Shifts tend to be considerably less for E modes than for F modes, and slightly less for oblique than vertical incidence.

Evidence from over-the-horizon radar[8] on single hop paths between 2000 km and 4000 km shows that quiet conditions usually prevail, since resolution down to 0.1 Hz is often possible. When the ionosphere is disturbed, however, such as occurs during conditions giving rise to spread F, there is typically a continuum of shifts of sometimes 5–10 Hz. During strong solar flares, deviations of up to 50 Hz have been measured[6] but these are only caused for a matter of minutes and are most unusual. Typical shifts caused by flares are 1–2 Hz.

Since each mode of propagation is composed of a number of rays which traverse slightly different ray trajectories, each ray path has a slightly different frequency shift. This results in a spread of received frequencies. Measurements[9] have not, unfortunately, been concerned with the shift and spread of individual modes; they have recorded composite Doppler values involving many modes. Under quiet conditions, spreads of 0.02 Hz would be applicable to E modes and 0.15 Hz to F modes[7].

Continuous Doppler spread modulates each transmitted pulse and therefore contributes to the fading of the received pulses. The fading period has been found[10], however, to be much greater than typical pulse durations.

4.4.5   Delay Distortion

Delay distortion occurs because the group delay is a function of frequency and is consequently not constant across a signal bandwidth. For a given ionospheric path, the oblique ionogram gives the frequency–time dispersion characteristics. Two examples are shown in Figure 4.21. The dispersion caused by the E layer is very small; the rate of change of group delay with frequency[11] is typically 5×10^{-6} µs/Hz. The F layer, particularly near its MUF, can cause much more rapid changes of group delay with frequency, as is evident from Figure 4.21.

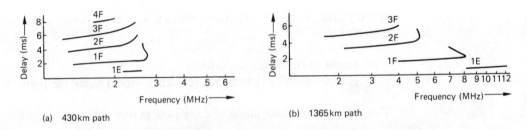

(a)   430 km path   (b)   1365 km path

Fig. 4.21   *Ionograms showing dispersion characteristics*

The importance of delay distortion for data transmission is concerned with the rate of change of delay with frequency and time. Ionospheric channels are non-stationary in both frequency and time, but if consideration is restricted to band-limited channels (say 10 kHz) and sufficiently short time (say 10 minutes) most channels are nearly stationary and can be represented adequately by a stationary model[7]. This means that, since propagation is limited to a discrete number of modes, the channel can be modelled by a delay line with a discrete number of taps, each of which is modulated in phase and amplitude by a time varying quantity.

4.5 Nearly Vertically Incident Sky Waves

4.5.1 The Need for Short-range Sky Wave Links

Chapter 3 has shown that terrain topography has an important effect upon the range achievements using the ground wave. Although radio waves in the HF band can be diffracted around obstacles and small hills, large obstacles can reduce the received HF field strength considerably. Thus, for example, large mountains in Norway effectively eliminate HF ground wave communication between fjords. Vegetation and ground conductivity can also affect ground wave range markedly; range over sea water is considerably better than over land.

For many short-range links HF ground wave propagation cannot provide an adequate service; nearly vertically incident sky wave (NVIS) propagation may be a better solution. For overland transmission paths the sky wave may be necessary owing to mountainous or heavily forested terrain or poor soil conditions as in deserts or arctic regions. Ground wave propagation is much more effective over sea paths but the need for high angle sky wave propagation may still arise at longer ranges or when mobile communications are required close to deeply indented mountainous coastlines.

For the NVIS mode, energy is directed vertically upwards towards the ionosphere and returned to the surface of the Earth. Umbrella-type coverage is produced that is terrain independent. HF systems with appropriate antennas have the potential of providing communications coverage out to ranges greater than 50 km over any type of terrain. Once the fundamental path loss and ambient noise factors are overcome by the transmission system in a given geographical area, successful NVIS communications can generally be expected between two or more points out to a separation of approximately 300 km with complete independence from terrain features.

Use was made of the NVIS principle for ground-to-ground communications by the armed forces during the Second World War when transmission difficulties were encountered with VHF equipment operating in the jungles and mountainous terrain[12]. The principle was further pursued and used during the Vietnam conflict[13].

4.5.2 Selecting the Frequency

The choice of frequencies to achieve these short-range sky wave paths is different from those normally used for a long-range application. The best frequency of operation lies within a small pass band (sometimes only one or two MHz wide) within the 2 to 10 MHz frequency band. The position of this pass band or 'window' is a function of time of day, season of year, sunspot activity, geographical location, etc. Antenna polarisation and directionality are also important considerations when using the NVIS mode for communication. A horizontally polarised antenna is preferable, with higher gain at radiation angles near the zenith. An antenna providing a significant vertical component would enhance the close-in signal performance by ground wave propagation.

During periods of high sunspot activity in the daytime or for anomalous conditions such as sporadic E, frequencies in the region of 10 MHz may be used. Operation during periods of low sunspot activity is restricted to frequencies of 6 MHz and below; indeed for some winter night-time conditions of low sunspot number the MUF is actually less than 2 MHz. The frequency choice within the confines of the prevailing window may be further restricted by possible ground wave interference at short ranges or by other sources of interference. Once a 'suitable' frequency has been established however, the propagation characteristics are generally such that a satisfactory signal strength can be achieved. Deviating from this suitable frequency by approximately 1 MHz may have the effect of reducing the received field strength by some 10 dB or even losing the signal altogether.

The likely effect as far as NVIS is concerned on onset of sporadic E conditions is a rapid and considerable increase in the available frequency 'window'. However, in order to capitalise upon this phenomenon, the communicator must be aware of the presence of sporadic E, which implies that some real time sounding facility is available (see Chapter 8).

Ionospheric storms may also cause a radical change in frequency window. Their most prominent features are a reduction in f_oF2 and an increase in D-region absorption. The practical consequence of this lowering of the maximum usable frequency and increase in the lowest usable frequency is a narrowing of the usable frequency window. The effects of a magnetic storm on f_oF2 at a given location depend, in an involved way, upon the local time, season, magnetic latitude and storm time[2]. The most severe depressions in electron density, and hence reductions in f_oF2, occur about 24 hours after the onset of the storm in geomagnetic latitudes of 45° and higher. In intermediate latitudes only a mild depression occurs, and in equatorial regions an actual increase may be observed.

4.5.3   Signal Strength

Typical sky wave signal strengths appropriate to central Europe produced at a receiver 200 km from a 1 kW isotropic antenna are shown in Figure 4.22. Clearly it is preferable to use as high a frequency as possible consistent with ionospheric support in order to provide the optimum signal strength. Moreover, the higher frequencies also incur lower noise levels (but not necessarily interference levels) and tend to be more efficiently generated at the transmitting antenna. From Figure 4.22 it is apparent that the range of field strengths received from a 1 kW transmitter are between 25 and 40 dBμV/m. By choosing appropriate frequencies near the MUF, according to time of day, this range can be decreased to between 30 and 40 dBμV/m.

4.5.4   Dependence upon Range

For the NVIS mode, communication ranges are such that the curvature of the Earth and ionosphere can be neglected. The MUF is then given by (compare equation (4.5)):

$$MUF = f_o F2 . \sec \phi \qquad (4.15)$$

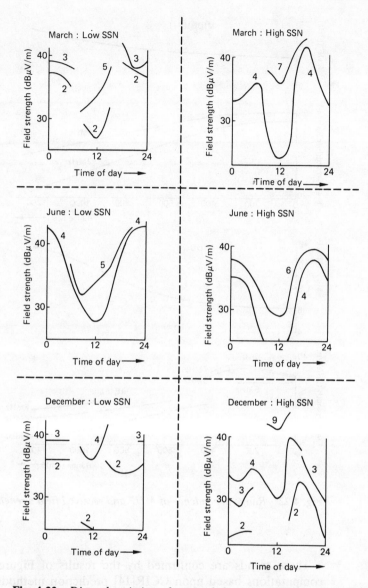

Fig. 4.22 *Diurnal variation of received sky wave field strength at 200 km from a 1 kW transmitter*

where φ is the angle of incidence at the reflecting layer (90° − elevation angle for a flat Earth). For a 10% deviation of MUF from f_oF2, φ = 25° corresponding to a range of approximately 270 km for the F2 layer reflections. Thus, the monthly median MUF should remain within about 10% of that for the vertically incident case up to that range. At greater distances the MUF deviates from equation (4.15) since the curvature of the Earth begins to become important. In the case of signal strength variations a similar argument can be applied. It can be assumed that, to a first approximation at small φ, the path length increases by a factor sec φ at oblique incidence and hence the received power decreases by a factor sec² φ.

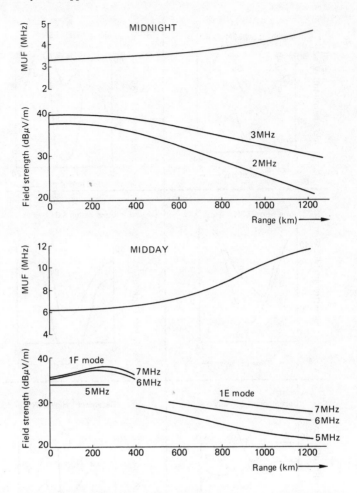

Fig. 4.23   *Range dependence of MUF and received field strength*

These trends are confirmed by the results of Figure 4.23 which shows computations based upon CCIR[14] prediction methods. Values given are appropriate for central Europe during March of a low sunspot number year. It can be seen from these values that up to ranges of about 400 km there is no significant change in MUF and the field strength changes by only 2 dB at most. At greater ranges, the MUF begins to increase more rapidly and the received strength decreases. In particular, during the day there may be a transition from F2 layer reflections to E layer reflections resulting in a sudden decrease of signal strength.

Figure 4.23 demonstrates that it is perfectly valid to extrapolate vertical incidence results when ranges of up to 400 km are of interest. For ranges greater than about 400 km, Figure 4.23 shows that the MUF begins to increase significantly with range and can no longer be considered similar to the vertical incidence value. Field strengths for a given frequency begin to decrease as the range increases. In the case of night-time conditions when the communications path is governed exclusively by the F2 layer, this decrease should be smooth. However, when there is the possibility of E and F modes being present simultaneously (as is usual for middle-of-the-day conditions) then transition may be abrupt as the dominant mode changes from an F2 mode (at near vertical incidence) to an E mode (at longer ranges).

4.5.5 Mode Structure and Multipath

The geometry imposed by NVIS operations means that E and F modes can occur simultaneously only for a very narrow range of frequencies close to the E layer critical frequency, f_oE. Under such conditions the wave partially penetrates the E layer (and is ultimately reflected from the F layer causing an F mode) and is partially reflected from the E layer (resulting in an E mode). If frequencies close to the E layer critical frequency are avoided, such multipath effects can be eliminated and only F modes or E modes, but not both, need be considered as constituting the received signal.

The difference in field strength between a one-hop and two-hop mode for a given layer is approximately the same as between a two-hop and a three-hop mode at short ranges. For F modes the difference is typically 15–18 dB and for E modes 20–25 dB (deduced from a sample of CCIR[14] predictions). Thus the multipath problems commonly experienced at longer ranges when, for example, the 2E mode is approximately equal in strength to the 1F2 mode tend not to be present at near vertical incidence.

Thus the major multipath effects are likely to be those caused by the interference of a sky wave mode with the ground wave. At ranges where these two modes occur with similar field strengths, strong interference may be expected. Discrimination against one or other of these modes may be effected by appropriate polarisation of the antenna, as mentioned in section 2.3.1.

4.5.6 Wave Polarisation

When an upgoing wave is incident upon the ionosphere it leads to the excitation of both the Ordinary and the Extraordinary wave. These two waves have different polarisations; they may be regarded as propagating independently within the ionosphere and are subject to different amounts of absorption[3]. For minimum path attenuation care should be taken to use a transmitting antenna whose polarisation leads to the excitation of a strong Ordinary wave and a receiving antenna polarisation corresponding to the larger resolved linear component of the downcoming Ordinary wave.

The antenna polarisation is not of critical importance for NVIS in so far as the polarisation coupling losses are concerned. Far more important is the need to discriminate against the ground wave. This can be achieved by avoiding the use of vertical polarisation. The advantage of using horizontal polarisation may also be apparent in terms of interference discrimination. For short-range transmissions on lower frequencies it is often convenient to use polarisation as a method of discriminating against noise. Very considerable gains in signal-to-noise ratio are possible in the daytime by using antennas which are insensitive to vertical polarisation when most of the interference is propagated via the ionosphere.

4.5.7   Summary of Attributes

A number of key points regarding the characteristics of NVIS communications can be summarised as follows:

a) The F2 layer within the ionosphere is of greatest importance.

b) The MUF at vertical incidence remains below 6 MHz during periods of low sunspot number but exceeds 6 MHz for 70% of the time during high sunspot number periods. Typical MUFs for vertical incidence are given in Table 4.2.

c) Operating frequency changes are generally only necessary at periods around dawn and dusk.

d) The MUF has its greatest diurnal variability in winter, its least in summer.

e) The range of field strength experienced over a wide variety of conditions only spans 10 dB to 15 dB when operating near the optimum working frequency.

f) MUF and signal strength are approximately independent of range for communication ranges less than 400 km.

g) Multipath problems experienced at longer ranges when two sky wave modes have approximately the same signal strength tend not to be present at these short ranges. Interference between sky wave and ground wave is the most likely multimode problem.

h) Antenna polarisation is not of critical importance in central European latitudes as far as polarisation coupling losses are concerned.

j) Sporadic E effects can widen the available frequency window considerably, but the occurrence of the phenomenon is extremely difficult to predict. Ionospheric storms have the effect of reducing the available frequency window.

5 Noise and Interference

5.1 Noise

5.1.1 Sources

Noise is the limiting factor which determines for every communications system whether or not the signal is usable for the transmission of information. It is very important to have a knowledge of the ambient noise with which the desired signal must compete so that sufficient power may be transmitted in order to overcome the effects of noise.

Electromagnetic noise impinging upon the Earth from solar and cosmic sources establishes a limit on the information that can be conveyed throughout the HF radio spectrum. This bound is modified by radiation from noise sources within the troposphere, the terrestrial environment and, primarily, from man-made radio frequency sources. Long-term characteristics of this noise affect the required transmission power, whilst short-term characteristics determine how the signal should be designed and detected to convey the desired information. In the HF and low VHF band, ambient electromagnetic noise fields are the dominant noise sources; noise from the receiver itself should be unimportant, as explained in section 5.1.7.

Although many types of noise sources may be present, the major contributing sources can be identified, for the HF band, as emanating from remote sources of atmospheric, galactic or man-made origin. The relative importance of each source varies as the radio frequency of interest is changed, and each is considered in the following sections. Local sources of electromagnetic noise can also be important under certain circumstances, as described later in section 5.3.

5.1.2 Noise Power

Thermal noise[1] is caused by the thermal agitation of electrons in conductors. It forms an ultimate bound upon the minimum attainable noise level. The noise generated in a unit bandwidth by a thermal noise source at temperature T_0 is kT_0 where k is the Boltzmann constant. Noise power is measured in terms of a quantity F_a which is defined as the external noise power available from a lossless antenna. It is usually expressed in terms of decibels above kT_0. The character of external noise is generally impulsive and non-Gaussian, but it is convenient to combine all of the contributions and define F_a as an 'effective' antenna noise power factor to provide a measure of the long-term rms value of the noise intensity. Typical values for F_a are shown for a variety of noise sources and different frequencies in Figure 5.1.

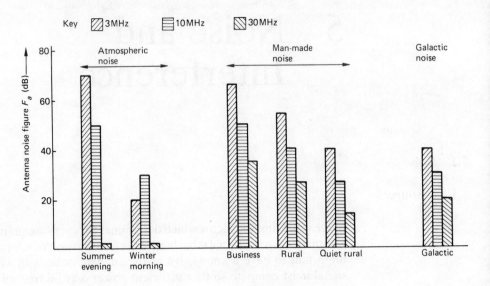

Fig. 5.1 *Effective antenna noise powers for a range of noise souces*

On the assumption that the noise power is proportional to the bandwidth b, in Hz, the total power P_N in dBW available at the terminals of a lossless antenna is given by[2]

$$P_N = F_a + B - 204 \qquad (5.1)$$

where $B = 10\log_{10}b$ and $10\log_{10}kT_0 = -204$ dBW, when T_0 is taken to be 290K.

It is often convenient to express the noise power in terms of E_N, the rms noise field strength. If E_N is expressed in dB above 1 μV/m, then

$$E_N = F_a + 20\log_{10}f_M - 95 + B \qquad (5.2)$$

where f_M is the frequency in MHz.

5.1.3   Atmospheric Noise

Atmospheric noise is produced mostly by lightning discharges in thunderstorms. The noise level thus depends upon radio frequency, time of day, weather conditions, season of the year and geographical location.

In the HF band, atmospheric noise is the most erratic of the three major types of noise. It is generally characterised by short pulses with random recurrence superimposed upon a background of random noise. Averaging these short noise pulses over several minutes yields values that are nearly constant during a given hour. Since atmospheric radio noise levels are well correlated over relatively wide areas the data can be represented by worldwide contour maps. Subject to variations caused by local stormy areas, atmospheric noise generally decreases with increasing latitude. Noise is particularly severe during the rainy season in areas such as the Caribbean, East Indies and Equatorial Africa.

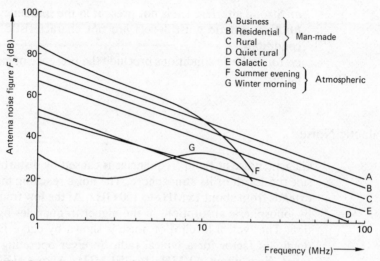

Fig. 5.2 *Antenna noise figure dependence upon frequency*

Atmospheric noise usually predominates in quiet locations at frequencies below about 20 MHz. Extensive measured data for ground sites is available[3]. For a particular locality there may be large daily and seasonal variations of noise intensity, amounting to some 40 dB variation in mean value at 3 MHz. To demonstrate this variability, mean noise levels for a near worst case of summer evening (curve F) and a near best case of winter morning (curve G) are shown in Figure 5.2.

The frequency dependence of atmospheric noise for the same season and time block presented by the CCIR[3] gives the distribution of radio noise throughout the world. Expected values of radio noise are based upon measurements made with a short vertical antenna over a perfectly conducting ground. The use of directive receiving antennas may modify the level of noise received. Equation (5.1) assumes that the noise is incident on the antenna uniformly from all directions.

5.1.4 Man-made Noise

The amplitude of man-made noise decreases with increasing frequency; it varies considerably with location. The noise arises chiefly from electric motors, neon signs, power lines and ignition systems located within a few hundred metres of the receiving antenna. Propagation commonly occurs by transmission over power lines and by ground wave. For frequencies below about 20 MHz, however, propagation may also be via sky wave reflection.

Measured data[4] from ground sites has enabled four typical levels to be defined, as shown by curves A to D in Figure 5.2. Short-term variations about these mean levels are large, with standard deviations in the region of 7 dB. Man-made noise may predominate under conditions of low atmospheric noise.

Curve D of Figure 5.2 is representative of the total external noise levels in the lower part of the HF band in quiet rural areas if

a) Strong interference is not present in the frequency band of interest.
b) Atmospheric noise levels are not characteristic of 'summer evening' conditions.
c) Ionospheric conditions preclude the presence of galactic noise in the HF band.

5.1.5   Galactic Noise

Galactic noise at radio frequencies is caused by disturbances originating outside the Earth or its atmosphere. The noise reaching the surface of the Earth extends from about 15 MHz to 100 GHz. At the low frequency end it is limited by ionospheric absorption, at the higher frequencies by atmospheric absorption. The typical level[5] of noise is shown by curve E in Figure 5.2. It is a dominant factor for a typical radio receiver operating within the frequency range from about 40 MHz to 250 MHz. Above about 250 MHz, internal receiver noise predominates. Below 40 MHz, other noise sources usually predominate. However, in quiet, remote locations the noise level from man-made sources is usually below galactic noise levels in the frequency range above 10 MHz.

5.1.6   Statistical Variations

The statistical variation of mean noise power is usually expressed in terms of D_u and D_d, the upper and lower *decile values* of the day-to-day variability of noise power relative to the median. Since these two values are sometimes not equal, this implies that the noise power distribution is often skew. However, for many purposes it is sufficient to assume that the noise distribution is symmetric with decile values

$$D = \tfrac{1}{2}(D_u + D_d)$$

and indeed in many instances D_u is approximately equal to D_d.

Theory shows that for a normal distribution, the standard deviation is related to the decile values by

$$1.28\ \sigma_N = D \qquad\qquad (5.3)$$

Thus, a knowledge of D_u and D_d will provide a value for σ_N. The magnitudes of D_u and D_d can be conveniently discussed in terms of different noise sources:

a) Atmospheric noise   The decile values are not only dependent upon radio frequency, but also upon seasonal and hourly factors. The CCIR[3] details the values of D_u and D_d. Some examples have been extracted and are presented in Table 5.1.

b) Man-made noise   Representative values of the day-to-day variability of man-made noise are given in Table 5.2 for different receiver sites and radio frequencies. The upper and lower decile values of day-to-day variability are assumed to be equal.

FREQUENCY	3 MHz											
Universal Time(UT)	0–4		4–8		8–12		12–16		16–20		20–24	
Decile value	D_u	D_d	D_u	D_d	D_u	D_d	D_u	D_d	D_u	D_d	D_u	D_d
Winter	9	7	11	10	9	7	9	7	12	10	9	7
Spring	8	8	13	12	11	8	14	10	15	13	7	7
Summer	8	8	13	12	13	9	17	12	15	14	7	7
Autumn	9	8	13	12	12	8	13	9	13	12	8	7

FREQUENCY	10 MHz											
Universal Time (UT)	0–4		4–8		8–12		12–16		16–20		20–24	
Decile value	D_u	D_d	D_u	D_d	D_u	D_d	D_u	D_d	D_u	D_d	D_u	D_d
Winter	5	4	7	6	8	7	8	7	8	7	7	6
Spring	6	5	8	8	8	6	10	7	10	9	6	5
Summer	5	5	7	7	8	6	11	7	9	8	4	4
Autumn	6	5	7	7	8	7	9	7	8	7	6	5

Table 5.1   Values of average atmospheric noise power D_u, D_d, exceeded for 10% and 90% of the time (dB above or below the median)

NOISE TYPE	FREQUENCY (MHz)		
	<10	10–20	>20
Business	4	6	7
Residential	6	9	12
Rural and quiet rural	10	7	4

Table 5.2   Decile values of average man-made noise power (dB above or below the median)

5.1.7   Condition for External Noise Limitation

Consider an incoming signal of mean power S at the antenna and an incident noise power N_0 in a 1 Hz bandwidth. Let the (omni-directional) antenna have efficiency η at the frequency considered. After reception by the antenna the signal power is ηS, the noise power is ηN_0 and the signal-to-noise ratio remains S/N_0. At the receiver input other losses such as those caused by imperfect matching reduce the signal strength to $\eta' S$ and the noise power to $\eta' N_0$.

As the signal passes through the receiver, the receiver noise power N_r per Hz contributes to the total noise power and the final signal-to-noise ratio becomes

$$\eta'S/(\eta'N_0 + N_r) \tag{5.4}$$

If conditions are such that

$$N_r \ll \eta'N_0 \tag{5.5}$$

then the signal-to-noise ratio at the receiver output is effectively the same as at the antenna. The inequality (5.5) can be expressed in logarithmic units as

$$\eta' + F_a \gg F_r \tag{5.6}$$

where η' dB is the loss in the receiving system, F_a is the *effective antenna noise power factor* (in dB) and F_r is the *receiver noise figure* (in dB). Inequality (5.6) is the condition that the receiving system performance is limited by external noise. Provided that (5.6) is valid the receiving antenna efficiency is unimportant. The condition in (5.6) is almost always true, except when using a poorly matched receiver at the lowest frequencies in the HF band under conditions of low external noise.

5.2   Interference

5.2.1   Sources

Serious communications problems are frequently present as a result of the occupation of the HF band by multiple (legitimate) users; the problem is particularly acute at night when the lower part of the band is severely congested. The interference is produced by large numbers of transmitters distributed over a wide area; it reaches the receiver mainly via sky wave propagation. In these circumstances successful communications may depend upon finding windows in the frequency band.

The need for accurate knowledge of how interference varies with frequency, time, bandwidth, threshold level and geographical location is vital for use in conjunction with propagation measurements to aid both the short-term channel selection process and the long-term channel assignments.

Central and Western Europe are well known to be particularly congested. The problem is most acute at night when traffic is concentrated towards the lower end of the band. Experience has shown that system performance is often limited by prevailing interference levels rather than propagation vagaries. It is thus necessary to study how interference characteristics vary with a variety of parameters.

5.2.2   Magnitude of Effects

The effects caused by interfering signals from a remote transmitter operating in the same frequency band depend upon radio frequency, time of day, season of year and sunspot number. If the radio frequency is too high, the sky wave cannot be returned to Earth; if it is too low, ionospheric absorption is excessive.

Although interference may be produced by a large number of individual sources distributed over a wide area, the effects can be demonstrated by considering a reasonably powerful single source of radiation. In this simplified approach, it is not intended to consider any details of statistical interference distribution.

The predicted MUF of a typical 1600 km temperate latitude sky wave path (characteristic of central Europe) is shown in Figure 5.3. These curves give an indication of the highest frequency likely to be a source of interference at this range. Consider now a 10 kW transmitter radiating isotropically. The field strength produced at such a range can be predicted[6] and is shown for three frequencies (2 MHz, 6 MHz and 10 MHz) in Figure 5.4. For the wave frequency of interest, it is then a simple matter to determine if the sky wave field strength exceeds the ambient noise levels determined at Figure 5.2. This will give an indication as to whether interference from other transmissions is likely to be a cause of concern. Note that the field strengths of the lower frequencies are considerably greater at night. This implies that more severe interference conditions will be present at these frequencies at night – a feature commonly experienced by communicators in the HF band.

A similar evaluation can be done for a shorter range (400 km) link with a 1 kW transmitter radiating isotropically. This reveals a somewhat different distribution of interference against frequency.

	Day (0600–1800)			Night (1800–0600)		
	2 MHz	6 MHz	10 MHz	2 MHz	6 MHz	10 MHz
10 kW transmitter at long-range (1600 km)	20	100	90	56	82	36
1 kW transmitter at short-range (400 km)	95	82	24	100	45	0

Table 5.3   Percentage of time interference exceeds ambient noise levels for a quiet rural site

Table 5.3 shows the percentage of the time (averaged over hours, seasons and sunspot activity) for which the median levels of the two postulated interference sources predominate over typical median ambient noise levels. Note in particular the interference caused at night by the 2 MHz frequency at the short range and in the daytime by the 6 MHz and 10 MHz frequency at the long range. In many cases the interference levels may exceed the ambient noise levels by 20 dB or more.

5.2.3   Variation with Time and Frequency

The occurrence of HF interference (or alternatively the level of spectral occupancy) tends[7] to be at its highest at night and at dusk, less at dawn and at frequencies near the daytime maximum usable frequency, and least at frequencies well below the MUF during the day.

If a quiet channel is defined as one with an interference level below that measurable (corresponding to a level of less than 1 μV of signal induced into the antenna) then measurements[8] show that typically

a) The maximum number of quiet channels occurs around midday.

b) The temporal variability (over a few minutes) of the HF interference spectrum does not change significantly with time of day.

A selection from a choice of, say, 16 channels could provide a channel with an interference level of not greater than 10 dBμV during the day, and 25 dBμV at night. The interference level of such a channel is unlikely to vary considerably over a minute or so. This suggests that, for the majority of short transmissions, a change of channel should not prove necessary. Moreover, there is often considerable directionality[10] of interference implying that significant advantage could be obtained by the use of directional antennas.

TIME	OPTIMUM WORKING FREQUENCY USED?	CONGESTION		VOICE BAND AVAILABILITY	
		Mean	*Standard Deviation*	*Mean*	*Standard Deviation*
Dawn	Yes	0.06	0.02	0.30	0.13
Day	Yes	0.26	0.10	0.05	0.08
Day	No	0.09	0.11	0.60	0.21
Dusk	Yes	0.43	0.15	0.03	0.03
Night	Yes	0.46	0.06	0.00	0.00

Table 5.4   Example values of congestion and voice band availability

Statistics gathered on congestion and voice band availability over a period of one week[7] are summarised in Table 5.4. In this example *congestion* is defined as the probability of finding, at random, a 100 Hz bandwidth within a 50 kHz spectrum, for which the average interference level exceeds a defined threshold of −125 dBm. *Voice band availability* is defined as the probability of finding a 2.5 kHz spectral window for which the interference level (measured within contiguous 100 Hz bandwidths across that window) is always below the same threshold. The definition of voice band availability may be considered to be a

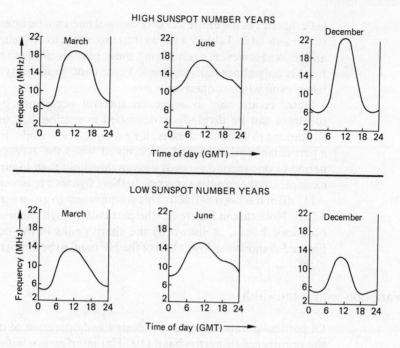

HIGH SUNSPOT NUMBER YEARS

LOW SUNSPOT NUMBER YEARS

Fig. 5.3   *MUF curves representative of a 1600 km sky wave path*

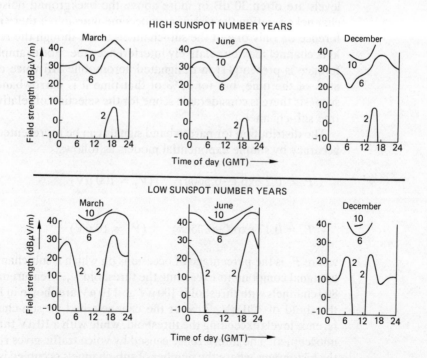

HIGH SUNSPOT NUMBER YEARS

LOW SUNSPOT NUMBER YEARS

Fig. 5.4   *Field strengths produced at a range of 1600 km by a 10 kW transmitter radiating isotropically (the numbers on the curves refer to the frequency in MHz)*

little 'harsh', as the entire 2.5 kHz channel need not be interference-free for use to be made of it. Table 5.4 shows the congestion to be consistently high at night and dusk. However, even during these periods of maximum congestion, the band is only about half occupied. Voice band availability fails to reach a high value even when congestion is low.

Statistics are also available on spectral occupancy by user[9]. The HF spectrum can be divided, as described in section 1.3, in terms of assigned allocations to designated users, for example aeromobile, fixed or broadcast. If a particular band is defined as 'occupied' when the average signal level over a period of one second exceeds a given threshold then Figures 5.5 and 5.6 show examples of the resulting spectra. In these figures a reasonably low threshold of -117 dBm has been defined. This is equivalent to an antenna noise figure F_a of 57 dB. Note that in Figure 5.5 the particularly high congestion is caused in the broadcast bands, as shown by the sharp peaks in the occupancy spectrum. Figure 5.6 shows the lower half of the HF band to be almost totally congested at night.

5.2.4   Variation with Bandwidth

Of particular importance to the design and operation of data transmissions is the occurrence of narrowband (100 Hz) interference within a 3 kHz channel. An examination[7] of the fine structure of HF spectral occupancy across frequency ranges of 50 kHz, using a resolution bandwidth of 100 Hz, shows that interference is often narrowband, relative to a 3 kHz channel, and that peak levels are often 30 dB or more above the background noise. For fixed sub-channel operation the circuit may become unusable if there is excessive interference on only one of the sub-channels, even though the remainder of the 3 kHz channel may be relatively interference-free. For example, average interference is present[11] in designated aeromobile HF voice channels for over 80% of the time, but for 25% of that time it is narrowband in nature. This suggests there is considerable scope for the selection of relatively interference-free sub-channels.

The distribution for narrowband signals can be represented with reasonable accuracy by simple exponential models as follows:

$$P_n = 0.2 \exp(-0.560n) \qquad (V_T = 100 \ \mu V) \tag{5.7}$$

$$P_n = 0.17 \exp(-0.356n) \qquad (V_T = 10 \ \mu V) \tag{5.8}$$

where P_n is the percentage of occasions on which n sub-channels are occupied by signal components exceeding the threshold V_T. Measurements[12] made on 800 channels with thresholds 100 μV and 10 μV are shown in Figure 5.7. With a threshold of 100 μV, 70% of the channels had no sub-channels with interference levels exceeding the threshold, while with a 10 μV threshold 51% were unoccupied. The interference caused by voice traffic gives rise to the peaks in the histograms where the number of sub-channels occupied is 16. These results tend to confirm that interfering signals in the HF band are essentially narrowband in nature.

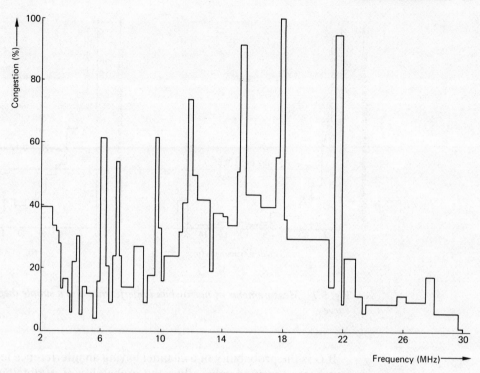

Fig. 5.5 *Daytime congestion values (−117dBm threshold) for July 1981*

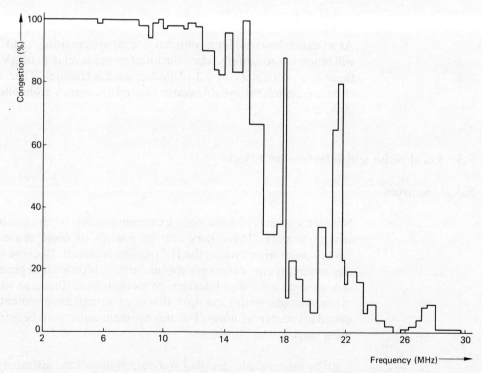

Fig. 5.6 *Night-time congestion values (−117dBm threshold) for July 1981*

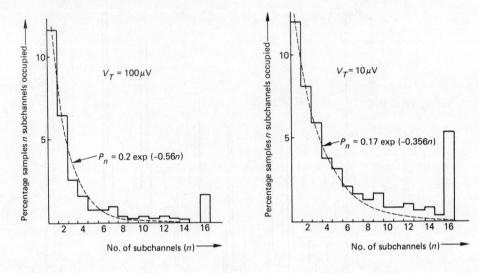

Fig. 5.7 *Measurements of narrowband interference and a simple theoretical model curve*

If G is the probability of a channel having an interference level less than or equal to a specified value, then the probability H of obtaining at least one success at receiving a signal after using m channels is given as

$$H = 1 - (1 - G)^m \qquad (5.9)$$

As an example, consider a communications system using a 3 kHz channel that will function successfully when the interference level is 10 dBμV. Suppose that there is a probability of 0.3 of finding such a channel. Then from equation (5.9), 6 channels (m) would need to be used to ensure a probability of success of 90%.

5.3 Local Noise and Interference Effects

5.3.1 Sources

Not all the effects of noise upon a communications system can be attributed to remote sources. There may also be sources of noise caused by the local environment within which the HF receiver is placed. Because individual local environments may differ very considerably, it is difficult to generalise a discussion of local noise and interference mechanisms. Recourse will therefore be made to an illustrative example, that of an aircraft environment, a particularly abundant source of noise. For this example noise may be generated by two main mechanisms:

a) by electrostatic charging and discharging of the airframe
b) by the rest of the aircraft electrical and electronic systems.

For *a*, noise is generated outside the fuselage and it must propagate through the atmosphere to the aircraft antenna; for *b*, noise is generated from within the fuselage.

5.3.2   Precipitation Static

Item *a* above is known as precipitation static. In unfavourable circumstances, and if no steps are taken to reduce it, the intensity of precipitation static noise can exceed the intensity of the noise from any other source in the HF band.

There are a number of mechanisms by which an aircraft can acquire or lose electric charge. Of the five main discharging mechanisms leakage current, corona and engine discharging cause a loss of charge from the fuselage as a whole; spark and streamer discharging merely cause a redistribution of charge between different parts of the fuselage.

All of the currents flowing in the charging and discharging processes radiate and contribute to the noise power received by an antenna on the aircraft. However, the noise power radiated by the rapidly fluctuating discharging currents is very much larger than that radiated by the comparatively smooth charging currents. Of the discharging mechanisms, corona, spark and streamer discharges are responsible for most of the precipitation static noise.

Propeller-driven aircraft can suffer corona discharge at the trailing edges of the propeller blades; a similar phenomenon has been observed on the rotor blades of helicopters. The corona discharging currents tend to make the noise occur in bursts.

5.3.3   Local Electromagnetic Interference

Aircraft-generated electromagnetic interference (EMI) is an extremely complicated subject. All aircraft electrical systems are likely to generate unwanted electromagnetic energy and this may couple into the aircraft radio systems and degrade their performance. For example, aircraft electronics systems generate various forms of EMI at HF including noise-like components; it is common experience that the noise received by the antenna increases as the aircraft systems are progressively switched on.

As measured at the antenna feeder, wide spreads of noise and interference levels appear to exist. These levels vary according to the type of aircraft and the electrical and electronic installations; there is also evidence of variations between individual aircraft of the same type.

Coupling mechanisms between aircraft installed systems and the antenna may be complicated; interference levels are affected by factors such as design, practice and workmanship of avionic installation, imperfect shielding of braided coaxial cables and the RF attenuation offered by the aircraft skin. With some aircraft the local noise fields are likely to be the largest single source of reception degradation.

Even for metal-skinned aircraft, there is some evidence[13] that there is non-uniform shielding at all frequencies and that resonances can occur at frequencies related to aircraft dimensions. As these frequencies tend to be in the HF band which contains some of the more powerful transmitters, this compounds the problem. The increasing use of composite materials exacerbates the problem still further. The aircraft structure and antenna configuration play a significant part in shaping the electromagnetic environment for an individual aircraft type.

Interference from other electrical and avionic systems may be of various types including impulsive, broadband noise-like and CW, depending on the nature of the source. Modern equipment is procured to approved electromagnetic emission standards (such as BS 3G 100[14]), but this does not necessarily assure interference-free HF reception. Coupling mechanisms to the HF antenna are not well understood and the level of coupling is dependent upon the individual installation.

5.3.4 Electromagnetic Compatibility

Emissions from an HF transmitter are frequently a cause of interference to other co-located electronic equipments. For example, transmissions from an aircraft HF radio may be a source of serious interference to other avionic systems. This is mainly as a result of the combination of high transmitter power and strong excitation of the whole airframe.

HF transmissions may generate very large electromagnetic fields in an aircraft. Since this is the near field régime the magnitudes and directions of the electric and magnetic field components are not related directly and each must be considered separately. For 100 W of power supplied, wire antennas may exhibit RF potentials in the region of 10 kV, producing fields of the order of 10 kV/m at the local aircraft surfaces.

Problems caused by EMI have become progressively more numerous and significant. As a result, there is increasing difficulty in achieving satisfactory electromagnetic compatibility (EMC) on aircraft, a trend which will undoubtedly continue as a result of increased avionic system complexity.

The attenuation and reflection properties of the materials used in an aircraft structure play a significant part in shaping the electromagnetic environment for an individual aircraft type. Fields may penetrate into the aircraft interior by various mechanisms. Direct penetration of the skin is negligible for a metal-skinned aircraft (skin depth is 0.07 mm at 2 MHz). However, fields may penetrate apertures in the skin and cause significant field strengths.

Fields may also penetrate through electrical imperfections in the skin such as imperfect hatch covers and joints; magnetic fields in particular may penetrate through narrow apertures. A modern trend in aircraft construction is to employ carbon fibre composite whose screening properties are poor for HF magnetic fields. Since more intense electromagnetic fields are present in the immediate vicinity of the antenna, particular attention must be paid to skin bonding in such areas.

Mechanisms of coupling between HF fields and avionic installations is a difficult and complex subject which is not understood fully.

5.3.5 Magnitude of Effects

Specification BS 3G 100 lays down maximum permissible radiated interference limits for equipment, expressed in dB relative to 1 μV/m. These limits can be more conveniently described in terms[15] of dB relative to antenna thermal

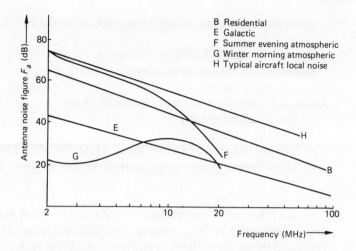

Fig. 5.8 *Comparison of levels of aircraft electrical noise with other noise sources*

noise power kT_0B for narrowband and broadband interference. The resulting values can, to a first approximation, be taken to represent the electromagnetic field levels existing inside the aircraft from all of the installed systems. The electromagnetic fields immediately outside the aircraft skin can be estimated by assuming a hull attenuation equal to 30 dB. If dischargers are fitted to the aircraft to reduce the effects of precipitation static, the antenna noise figures can be reduced by 40 dB to 60 dB. Under these circumstances the radiated interference, reduced by the appropriate value of hull attenuation, is the dominant noise source. This is shown in Figure 5.8 and can be compared with other sources of noise in the HF band.

5.4 Audio Noise

5.4.1 Sources

The problem of audio communications at HF is often exacerbated by the use of a poor-quality telephone system in a grossly noisy environment. The noise, being any unwanted sound that arrives at the listener's ear, may arise from a number of different origins.

The example of the aircraft as a difficult communications environment again serves to highlight some of the problems. Whilst electrical sources of audio noise are essentially the same for fixed-wing aircraft and helicopters the mechanisms by which each create acoustic noise are somewhat different. Fixed-wing jet aircraft create acoustic noise in their cabins from the flow of air over canopies, from air conditioning systems and from turbine power plants, roughly in that order of importance. Helicopters, however, create their noise from rotor-blade passage over the fuselage, from engine intakes and exhausts, from gearboxes and from other mechanisms.

Acoustic mechanisms involve transmission to the ear in one of two ways:

Directly (or through a helmet if worn)
Indirectly, by microphone pickup.

Audio noise can be generated electrically from:
a) Baseband interference
b) RF link interference
c) Intermodulation and other forms of distortion at baseband
d) Distortions through imperfections in the RF link, for example tuning drift.

Acoustic noise and shortcomings in the installed radio system have the effect of degrading speech both on transmission and reception. The problem is often enhanced by the use of poor operational practices, such as reducing perceived rms radio receiver noise to a level well below ambient acoustic noise.

5.4.2   Effects upon Speech Transmission

The quality of transmitted speech is affected by two mechanisms:

a) Noise entering the microphone
b) Modulation of speech production by sub-audio components of the ambient noise.

Usually the effects are not serious but large effects may sometimes occur, such as air-buffeting of the microphone in a helicopter when the cabin door is open.

5.4.3   Effects upon Speech Reception

Perceived audio signal-to-noise is degraded by ambient acoustic noise. Assessment of speech intelligibility in noise is a complex and difficult subject; it is not easy to translate perceived audio signal-to-noise ratio into terms of probability of successful communications. For assessment purposes it is necessary to define 'typical' overall degradation of reception due to ambient acoustic noise but, due to local circumstances, wide variations in these typical levels may occur in practice.

For satisfactory reception quality the signal has to be presented at a level such that it is audible above this noise. Increasing the signal level does not necessarily improve reception quality, however, since the ear itself cannot 'decode' into intelligible speech signals that are extremely loud. A significant number of messages may be misunderstood or perhaps missed altogether.

The effects of broadband noise on intelligibility cannot be assessed easily. Although broadband noise is difficult to prevent or reduce and has a major effect upon the intelligibility of speech, it may not be a major problem with the detection of signals which are 'tone-like', and is probably less annoying and fatiguing than interference of very narrow bandwidth.

6 System Performance Assessment

6.1 Antenna Considerations

6.1.1 Matching

Propagation effects and levels of ambient noise play a crucial role in HF communications link performance, as described in previous chapters. However, the signal and noise levels received over a given HF link can also depend considerably upon the antenna characteristics. Before an assessment of system performance can be undertaken, it is therefore necessary to review the role of the antenna and its impact upon HF link performance.

The antenna considerations posed by the high frequencies make communications problems complicated. The usefulness of system performance estimates depends on realistic power gain and radiation pattern estimates of the antennas on the operational circuit. In the assessment of a given HF link it is important to be able to pinpoint sources of antenna performance degradation.

To the communications engineer interested in the overall design of a radio communications system, the antenna is but one link in the complicated chain that leads from the message input to the message output. It is natural to consider the antenna simply as another circuit element that must be properly matched to the rest of the network for efficient power transfer. From this viewpoint the input or terminal *impedance* of the antenna is of primary concern.

Many HF antenna systems that are electrically short (i.e. their physical size is much smaller than a wavelength) exhibit low radiation resistance r and high reactance x. An antenna tuning unit (ATU) is required to tune out the reactive term and to present a suitable resistance to the transceiver. Both the antenna and the ATU exhibit loss resistance, which for the tuned and matched condition is expressed as an equivalent series loss resistance r_1. The antenna system *radiation efficiency* η is defined as the ratio of power radiated by the antenna to the power supplied by the transmitter thus:

$$\eta = r/(r + r_1) \tag{6.1}$$

6.1.2 Gain and Directivity

The *gain* of an antenna is a measure of how well the antenna concentrates its radiated power in a given direction. It is the ratio of the power radiated in a given direction to the power radiated in the same direction by a standard antenna (usually an isotropic antenna or a dipole), keeping the input power constant. If the pattern of the antenna is known and there are no ohmic losses in the system, the gain G is defined by

$$G = \text{maximum power intensity/average power intensity}$$

$$= 4\pi|E_0|^2 / \iint_\Omega |E|^2 \, d\Omega \tag{6.2}$$

where $|E_0|$ is the magnitude of the field at the maximum of the radiation pattern and $|E|$ is the magnitude of the field in any direction. Power gain is an inherent property of the antenna and does not involve impedance or polarisation mismatch losses. All actual antennas have radiation patterns such that the power flux is maximised in certain preferred directions. When the total power radiated by an actual antenna is the same as that for a hypothetical isotropic antenna then the gain in the maximum direction must be exactly balanced by a loss in other directions.

The *effective area A_r* of an antenna is defined by

$$A_r = G\lambda^2/4\pi \tag{6.3}$$

where G is the antenna gain and λ is the wavelength. The power delivered by a matched antenna to a matched load connected to its terminals is PA_r, where P is the power density in Wm^{-2} at the antenna and A_r is the effective area in m^2.

The effective area is thus the ratio of power available at the antenna terminals to the power per unit area of the appropriately polarised incident wave. For a lossy antenna the effective area determines the useful power delivered to the load, and hence indicates the efficiency.

For an HF circuit it is often the directivity of the antenna rather than its absolute gain that is important on receive. *Directivity* is the property of the antenna that enables it to discriminate in favour of signals coming from a particular direction. Directivity D and gain G of antennas are directly related in terms of the antenna efficiency η, thus

$$G = \eta D \tag{6.4}$$

Directionality arises from the way the radiation from various parts of the antenna combines with different phase in different directions. The radiating parts may be separate sections of the same antenna element (for example, in a long wire antenna or a transmitting loop) or they may be independent radiators (some of which may be parasitic). In addition the vertical pattern is influenced by reflections from the ground; these are usually taken into account by assuming the presence of an image of the antenna below ground level.

Directional antennas have a number of advantages including an improvement of signal-to-noise ratio, an increase in traffic handling capacity (by rotating the beam or by using more than one directional antenna) and a reduction of the susceptibility of the receiver to interference.

It is not uncommon to specify the same antennas for both reception and transmission in a communications link. This is a simple approach which is often justifiable. However, the antenna efficiency is often not of consequence for reception and for this reason it is important to give separate consideration to antennas used for reception. In some cases the most suitable type of antenna for reception is entirely different from the antenna used for transmission.

6.1.3 Polarisation

To obtain maximum transfer of power between two antennas, the antenna and the characteristic wave polarisation should match. In the process of propagation, however, the polarisation may change. For the ground wave this may be caused by surface reflections; for the sky wave passage through the ionosphere in the presence of a magnetic field is likely to impart elliptical polarisation to a plane polarised incident wave.

For ground wave propagation it is necessary to use vertically polarised antennas at both ends of a link, because horizontally polarised signals are attenuated very rapidly. With sky wave propagation the plane of the polarisation of the received signal is time-dependent and generally differs from the transmitted polarisation; this variation in polarisation is one of the factors contributing to short-term fading.

Thus if sky wave propagation only is concerned, the polarisation of the transmitting and receiving antennas can usually be chosen for convenience in antenna design. This is an important freedom, because it makes it much easier to design antennas with specified radiation patterns, especially in the presence of physical constraints. (The simultaneous use of receiving antennas with orthogonal polarisations for diversity reception is discussed in section 6.1.7 below.)

The loss on ground reflection depends upon the nature of the soil, being least for sea water. However, the loss also depends upon the angle of incidence and, at grazing incidence, the loss is very small even over soil of poor conductivity.

Horizontally polarised signals undergo a phase reversal on reflection. A consequence of this is that efficient low angle radiation with horizontal polarisation requires a high antenna, so that there is a half wavelength path difference between direct and reflected rays at low angles.

Vertically polarised signals undergo a phase reversal at angles of incidence below the so-called *Pseudo-Brewster angle* but not above this angle. The Pseudo-Brewster angle decreases with increasing soil conductivity, being least for salt water. In consequence the lowest effective radiation angle with vertical polarisation depends upon the nature of the soil. (The option of raising the antenna well above the ground is rarely practicable for a vertically polarised radiator at HF.)

6.1.4 Arrival Angles

The angle of arrival in azimuth of a sky wave signal can differ from the great circle bearing of the transmitter. Variations of about 1° rms can be regarded as typical, although larger variations are sometimes observed under disturbed ionospheric conditions. In addition, auroral effects can cause azimuthal arrival

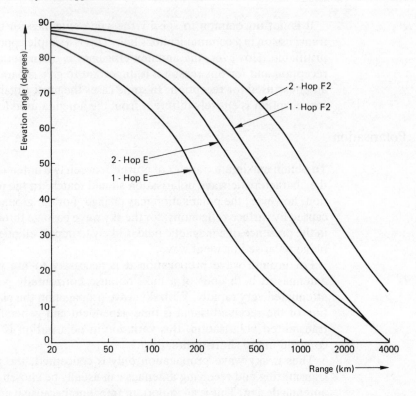

Fig. 6.1 *Elevation angles of sky wave modes against range for typical ionospheric layer heights*

angles to differ substantially (by tens of degrees in some cases) from the great circle bearing[1].

The elevation angle of radiation for a sky wave signal depends upon the great circle ground range and which ionospheric modes are present. In Figure 6.1 curves are given for ranges up to 4000 km, for typical virtual layer heights. At ranges greater than 4000 km the best propagation is generally obtained at the lowest radiation angles which can be accommodated in terms of antenna design[2].

It is important that transmitting and receiving antennas at the two ends of a link can couple into the same propagation modes. For example, it is of little value to provide a receiving antenna capable of receiving at very low angles, if the transmitting antenna at the other end of the link cannot radiate effectively at such low angles.

6.1.5   Transmitting Antennas

In order to provide maximum radiated power, a transmitting antenna is required to present to its drive source an impedance which remains inside certain limits, over the frequency range within which it is designed to operate. This is normally specified in terms of a nominal impedance (assumed to be resistive), and a maximum permissible voltage standing wave ratio (VSWR).

Typical values of impedance are 50 ohm, 75 ohm, 200 ohm, 300 ohm, and 600 ohm. Typical maximum VSWR values are 1.2:1, 1.5:1, 2:1, 2.5:1, and 3:1. Other values of impedance and maximum VSWR are sometimes specified.

In relation to matching, transmitting antennas for HF operation can be grouped broadly into three classes:

1   Self-resonant Antennas
These antennas, for example the half-wave dipole, operate at a specific frequency. Some types, for example the variable length monopole, can have their geometry changed to vary the resonant frequency.

2   Resonated Antennas
These antennas can be used over a range of frequencies, by employing a matching unit which transforms the widely varying antenna input impedance to some specified value. An example is the end-fed whip antenna.

3   Broadband Antennas
These antennas are designed to present an impedance which remains within an acceptable range over a wide band of frequencies. Some types, such as the fan dipole, are basically self-resonant antennas with a very broad bandwidth capability. Other types operate on different principles; an example is the rhombic, which is a travelling wave antenna.

Self-resonant antennas are appropriate only if one or two spot frequencies are ever used. If operation at many frequencies is required the ideal antenna is one of the broadband types, chosen to have an operating band covering the necessary range. Unfortunately broadband antennas at HF are large devices, their size being proportional to the longest operating wavelength; size also tends to increase as the required operating bandwidth is increased. If physical size rules out the use of a broadband antenna, then a resonated antenna must be used. This introduces considerations of matching unit losses and tuning times.

It is not usually acceptable to connect a broadband antenna directly to a broadband transmitter even if the VSWR of the former remains within the limits acceptable to the latter. Some tuning or sub-octave filtering is usually necessary, to reduce harmonic radiation.

6.1.6   Receiving Antennas

Many commercially available antennas can be obtained in receive-only versions; these differ from the transmitting versions in that no significant power handling capability is needed, which permits cheaper construction.

The efficiency of a receiving antenna, and of the means by which it is coupled to the receiver, need only be sufficiently high to ensure that the system is externally noise limited. In general, HF receiving systems should be designed to keep RF signals at the minimum level consistent with external noise limiting, at every point in the system, to minimise the levels of intermodulation products[3].

The fundamental requirement is to achieve the best signal-to-noise ratio in a receive antenna. Whereas in a transmitting antenna power gain is paramount, in the case of receiving antennas it is the directivity which is the important

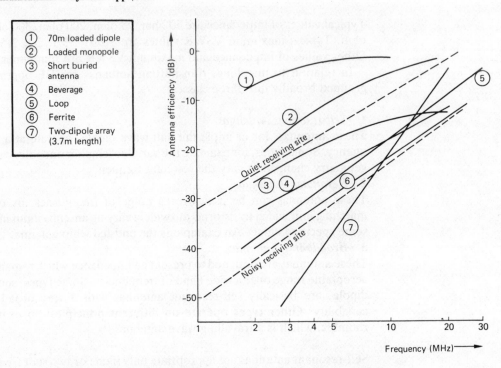

Fig. 6.2 *Examples of antenna efficiencies for some HF antennas*

parameter. Thus it is possible to use antennas of small size and to achieve directivity in ways which are not practicable for transmitting antennas, such as by the use of phased arrays of monopoles or loops.

Figure 6.2 shows typical antenna efficiencies against frequency for a number of different antenna types. A receiving antenna at a noisy receiving site needs only to have efficiencies exceeding those shown in the lower broken curve of Figure 6.2. These minimum efficiency values are met by loops, ferrite loops, Beverage and buried antennas. Thus all these four types come into consideration provided their polar diagrams are also suitable.

Antennas such as the horizontal dipole, the monopole, log periodics and rhombics all meet the efficiency criterion. However, where the receiving antenna is located at a very quiet receiving site most of the more inefficient antennas would be ruled out, since their noise levels would limit the performance under conditions of low atmospherics and ambient noise. This is illustrated in Figure 6.2 by the upper broken line. This shows the lowest permissible efficiency for the extreme case of a low reliability system, little atmospheric noise and quiet rural surroundings.

6.1.7   Diversity Operation

Fading of radio signals, especially the type referred to as 'fast fading', degrades the received signal intelligibility. However, the phenomenon of fading itself has characteristics which can be utilized to minimise its detrimental effects.

Diversity systems function on the basis of reception of two or more uncorrelated signals from the same source. A selection or combining scheme is applied to these signals to minimise the effects of fades. Thus fading effects caused by multipath (selective fading) can be reduced by diversity systems.

Three of the methods of providing multiple uncorrelated inputs involve antennas:

1 Spaced Antenna Diversity

With spaced antenna diversity, two antennas (or occasionally more) are used to provide signals for the receiving system. The antennas are usually, but not necessarily, identical types. The relationship between the antenna spacing and the correlation coefficient of short-term fading is given sufficiently accurately for practical purposes by the following expression:

$$C = \exp\left(-S^2/2S_0^2\right) \tag{6.5}$$

where C is the *correlation coefficient*

S is the antenna spacing

S_0 is the antenna spacing at which the correlation coefficient is reduced to 0.61.

There is some uncertainty at present regarding the values of S_0, and whether the value is different for spacings parallel to and perpendicular to the direction of propagation. The current consensus is that S_0 is around 10–15 wavelengths, and that the direction relative to the line of propagation is in general immaterial. However, if circumstances necessitate the use of spacings of less than about 150 m the antennas are best spaced along the direction of propagation[4].

2 Polarisation Diversity

This diversity method depends upon the simultaneous reception of the same transmission on antennas of two separate polarisations. Below 4 MHz this method may fail to work for an appreciable part of the daylight hours on account of only one magneto-ionic component being present. Results for higher frequencies during daytime, however, are more promising. For example, when error rates of between 1 in 10^2 to 1 in 10^3 are recorded separately for 10 MHz on vertically and horizontally polarised antennas, the combined signals can reduce this error rate by an order of magnitude.

Polarisation diversity appears[5] to be similar in effectiveness to spaced antenna diversity using two antennas at least 300 m apart. These results make polarisation diversity appear to be very attractive in comparison with space diversity because of the saving in real estate. However, there are practical problems to be considered. It may be necessary to use vertical polarisation to permit ground wave operation, or to use horizontal polarisation to assist in rejecting ground wave interference and man-made noise (which tends to be vertically polarised). Even if both polarisations are equally acceptable from the viewpoint of propagation there may be constraints associated with the required radiation pattern; some types of radiation pattern are difficult to obtain with one or other polarisation.

3   *Arrival-angle Diversity*

Arrival-angle diversity requires the use of a receiving antenna with discrimination in the vertical plane. For this reason it does not appear to be a method that is in common use for HF. With a vertically steerable array comprising short vertical monopoles or active loops, worthwhile improvements are possible by isolating the different ionospheric modes (assuming that they have sufficient angular separation) and using them in diversity. In many circumstances the mode separation could be performed adequately assuming mode arrival angles based on propagation forecasts, obviating the need for complex closed-loop methods of pattern control.

6.2   Signal-to-Noise Ratios

6.2.1   The Received Signal-to-Noise Ratio

The received available signal power S at the output of the receiving antenna terminals can be expressed in decibels relative to 1 watt as follows:

$$S = P_t + \eta_t + D_t - L + \eta_r + D_r \tag{6.6}$$

assuming no polarisation mis-match loss, where

P_t = transmitter power (in dBW)
L = path loss in the given direction (in dB)
η_t, η_r = transmitting and receiving antenna efficiencies (in dB)
D_t, D_r = transmitting and receiving antenna directivities for the signal source (in dB).

The absolute gains (again in logarithmic units) of the antennas with respect to isotropic antennas are

$$G_t = \eta_t + D_t$$
$$G_r = \eta_r + D_r \tag{6.7}$$

The received noise (in units of dBW) in the receiver bandwidth B is

$$N = kT + B + F_a + \eta_r + D_n \tag{6.8}$$

where kT = thermal noise power density (-204 dBW per Hz)
B = receiver bandwidth (in dBHz)
F_a = effective antenna noise power factor which results from external noise power available from a loss-free antenna (in dB)
D_n = receiving antenna directivity factor for the sources of noise (in dB).

Thus the received signal-to-noise ratio S/N (in dB) in the bandwidth B is, from equations (6.6) and (6.8),

$$\text{S/N} = S - N = E_p - L + D_r' - N_i \tag{6.9}$$

where E_p = effective radiated power

$D'_r = D_r - D_n$ = receiving antenna directivity factor against far field noise

$N_i = kT + B + F_a$ = noise power incident at the input terminals of the antenna.

The signal-to-noise ratio in a 1 Hz bandwidth is

$$S/N_0 = S/N + B \qquad\qquad (6.10)$$

6.2.2 The Required Signal-to-Noise Ratio

To assess the system performance it is first necessary to adopt a minimum required S/N_0 ratio criterion for a given grade of service. For example the CCIR[6] recommend a 15 dB S/N ratio for marginal and 33 dB for good commercial quality of communications with HF SSB voice in a 3 kHz bandwidth. This corresponds to a S/N_0 ratio, see equation (6.10), of 50 dB for marginal and 68 dB for good commercial quality. In practice criteria must be chosen to suit the communications requirements. For example, higher S/N_0 ratios than the minimum usually specified might be required to reduce data errors due to bursts of noise or to achieve intelligibility of unrelated words on speech transmission.

6.2.3 The Median Signal-to-Noise Ratio

The median signal-to-noise ratio can be derived from numerous individual measurements of the instantaneous signal-to-noise ratio. A single measurement of grade of service between two locations at a specified time implies a single value of the signal-to-noise ratio. Further measurements at different times or at different locations (at the same range) will yield some statistical spread of values in the signal-to-noise ratio.

From a knowledge of the variability of the signal and of the noise it is then possible to determine the probability of exceeding some specified required signal-to-noise ratio. This, in turn, provides a percentage success figure for achieving a given error rate or grade of service.

The monthly median value of the mean signal power S, for a particular propagation mode, may be combined with the monthly median of the mean noise power N_0, within a 1 Hz bandwidth, to produce an estimate of the monthly median value of S/N_0, the signal-to-noise ratio in a 1 Hz bandwidth. From a knowledge of the day-to-day variability of the signal and the noise it is then possible to determine the probability of exceeding some specified required S/N_0 ratio.

6.2.4 Variability of Signal-to-Noise Ratio

Let P be the percentage probability that the mean received S/N_0, whose monthly median is R_m is above some specified level R_0.

If the statistical variation of received signal-to-noise ratio follows a Normal Distribution, then the probability of communications success will be governed by two parameters:

a) The difference (in dB) between the median received S/N_0 ratio, R_m, and the required S/N_0 ratio, R_0, for the desired grade of service.

b) The standard deviation σ_T, which characterises the spread of S/N_0 ratio values about the median.

Let $t = (R - R_0)/\sigma_T$

$\qquad x = (R_m - R_0)/\sigma_T$ (6.11)

where R is the instantaneous value of S/N_0 ratio, R_m is the median value and R_0 is the minimum desired value for communications 'success'.

Then the probability of R exceeding R_0 is

$$A(x) = \int_{-\infty}^{x} (2\pi)^{-\frac{1}{2}} \exp\left(-\tfrac{1}{2}t^2\right) \; \mathrm{d}t \qquad (6.12)$$

Thus $A(x)$ is the probability of communications success for a median received S/N_0 value of R_m and a desired value of R_0. Equation (6.12) is plotted in Figure 6.3. In order to make use of this curve, some value must be assigned to σ_T.

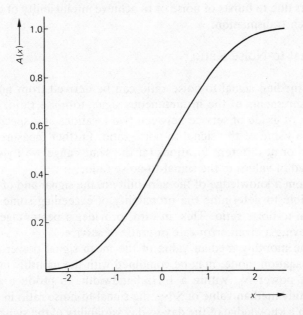

Fig. 6.3 *Probability distribution for communications 'success' [equation (6.12)]*

The value of σ_T can be derived in terms of σ_S and σ_N the standard deviations for the signal and for the noise respectively. Assume that each variation has a normal distribution and let the variance of the mean signal level be σ_S^2 and the variance of the mean noise level be σ_N^2. Then the central limit theorem shows that variations in signal-to-noise ratio are governed by a normal distribution of variance:

$$\sigma_T^2 = \sigma_S^2 + \sigma_N^2 \qquad (6.13)$$

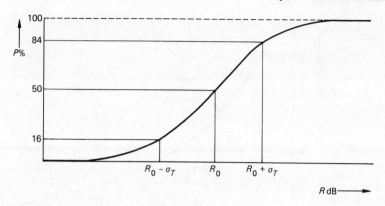

Fig. 6.4 *Probability of S/N ratio exceeding R_0 if R is the instantaneous received value of S/N ratio*

Figure 6.4 shows the probability P of the instantaneous signal-to-noise ratio, whose median value is R_m exceeding some required value R_0. Clearly, the larger the value of σ_T the greater the spread of signal-to-noise and the larger the value of R required to ensure a high probability of communications success. The value of P is a more convenient way, as will be seen later, of describing the same information as $A(x)$.

6.3   Circuit Performance Criteria

6.3.1   Ground Wave Variability

The statistics of the received signal-to-noise ratio can be used to define some circuit performance criteria. For ground wave transmissions the two major statistical parameters of interest are the variability of signal strength caused by local station siting conditions and the noise variability due to random fluctuations in noise emissions. A value of σ_S equal to 6 dB seems to be appropriate for frequencies between 20 MHz and 70 MHz but at lower frequencies siting does not appear to be critical, except for obviously very bad sites. The variation of signal-to-noise ratio in the HF band for ground wave transmissions is therefore almost wholly caused by noise statistics. The decile variations for various noise sources have already been described in Chapter 5. If D is the difference in dB between the median and decile value of noise level then D can be used to assess the circuit performance. A simple example is shown in Figure 6.5 which gives values for 50% confidence of achieving a signal-to-noise ratio R_m dB at a given range. The value of signal-to-noise ratio achieved with 90% confidence at that same range will be $R_m - D$. Thus, for example, if 50 dB S/N$_0$ is required for a given grade of service and $D = 10$ dB, then Figure 6.5 shows that a range of 80 km can be achieved with 50% confidence but the range is reduced to 50 km if a 90% confidence is required.

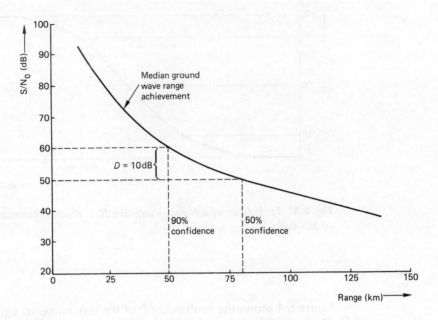

Fig. 6.5 *Effect of noise variability upon ground wave range confidence values*

6.3.2 Sky Wave Variability

The simple example given above for ground wave circuits cannot be applied directly for the sky wave since a further factor, namely the probability of ionospheric reflection, must also be included.

The transmission loss of an HF sky wave propagation path depends upon a number of factors, and these can have implications on various aspects of the communications link. Day-to-day variations in the electron concentration within the E- and F-regions of the ionosphere influence the direction which a given ray path takes through the ionosphere to reach the receiver. These fluctuations in ray path direction may, in turn, change the effective gains of the transmitting and receiving antennas. Different ray path directions can produce different effects of focusing, spatial attenuation, sporadic E losses, polarisation losses and multipath phenomena. Changes of electron concentration in the D- and E-regions can affect considerably the value of ionospheric absorption. Day-to-day variations may also be produced in atmospheric noise intensities by these mechanisms.

These day-to-day variations of signal strength have been examined for a wide range of paths and operating conditions. The greatest variations occur for paths in the range of 65° to 70° geomagnetic latitude.

In contrast to the ground wave, therefore, it is the signal variations that often exceed those of the ambient noise. A knowledge of these distributions as characterised by σ_S and σ_N for signal and noise respectively can provide a value for $A(x)$, defined in equation (6.12), and hence for P in Figure 6.4. However, P alone does not characterise a sky wave circuit satisfactorily; a further statistical parameter, that of sky wave availability, must be introduced.

6.3.3 Sky Wave Availability

The upper limiting frequency of sky wave propagation is governed by the distribution of electron concentration along the propagation path. This upper limit is known as the maximum usable frequency (MUF) and may be considered as the maximum frequency of a wave capable of propagating over a given sky wave path. Because the properties of the ionosphere exhibit temporal variations, the MUF fluctuates continually; but predictions of ionospheric conditions are based upon monthly median conditions and some allowance must be made for day-to-day variability. Thus the predicted MUF is defined as that frequency for which signals are expected to be available for 50% of the days at a given hour within a given period, usually a month (see Figure 6.6). A note of caution should be made. The MUF concept can only be considered an approximation since many factors can lead to errors in estimates of the MUF. For example, when the ionosphere is highly structured spatially the HF signal may not disappear completely when the frequency is above 'the MUF'.

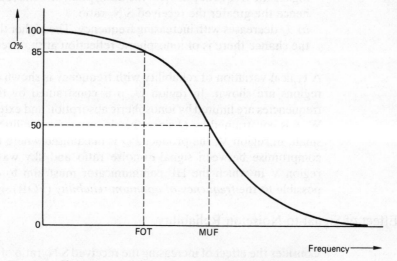

Fig. 6.6 *Sky wave availability Q as a function of frequency*

The *sky wave availability Q* is defined as the percentage probability that radio signals can propagate at a given hour over a given sky wave path. It may be derived in terms of the statistics of the day-to-day variations of the MUF. The local MUF may be increased significantly by the presence of sporadic E ionisation.

6.3.4 Circuit Reliability Factor

The *circuit reliability factor* is defined[7] as the fraction of days that successful communications may be expected at a given hour within the month at a specific operating frequency. It is based upon monthly median estimates of propagation parameters and their distributions and represents the fraction of days in the month at the given hour that successful communications can be expected.

The primary factor in determining the circuit reliability is the long term median S/N_0 ratio. This is directly associated with a grade of service, which in turn defines the type of communications desired; for example, the percentage of error-free messages in teletype transmissions or the intelligibility of voice transmissions. A minimum required S/N_0 ratio is associated with the desired grade of service. This ratio depends upon many factors such as modulation index, signalling rates and codes and includes effects of fading, error-correcting schemes, optimum modulation and detection techniques and diversity schemes.

Mathematically, the circuit reliability factor can be expressed as a dimensionless quantity given by

$$\rho = PQ \qquad\qquad\qquad (6.14)$$

The magnitude of ρ depends ultimately upon two opposing effects:

a) P increases with increasing frequency. For a given sky wave path the higher the frequency the less the absorption and the less the received noise; hence the greater the received S/N_0 ratio.

b) Q decreases with increasing frequency. The higher the frequency the less the chance there is of ionospheric reflection at that frequency.

A typical variation of reliability with frequency is shown in Figure 6.7. Three regions are shown. In region U, ρ is constrained by the value of P since frequencies are limited by ionospheric absorption and external noise; in region W, ρ is constrained by Q since the sky wave availabilities of frequencies are small; in region V, the product PQ is maximised where there is an optimum compromise between signal-to-noise ratio and sky wave availability. It is region V in which the HF communicator must aim to operate, as close as possible to the *frequency of optimum reliability* (FOR).

6.3.5   Effect of Signal-to-Noise on Reliability

Consider the effect of increasing the received S/N_0 ratio of the HF link by some means such as increase of transmitter power, improvement of antenna efficiencies or the reduction of received noise. Then P increases, as shown in Figure 6.8 by curves A to C, and hence ρ also increases (curves A' to C' in Figure 6.9). The factor Q, however, remains constant and shows the maximum value of ρ which is attainable. As Figure 6.9 shows, the Q curve forms an 'envelope' to the reliability curves.

Thus, for example, if the median received S/N_0 ratio R_m is such that curve C' is applicable in Figure 6.9 and if 85% reliability is required, then a horizontal line drawn from $\rho = 85\%$ cuts curve C' at two values of f, denoted by f_1 and f_2. The lower value f_1 is known as the lowest usable frequency, LUF, and the upper value f_2 is just below the optimum traffic frequency, FOT. For frequencies below the LUF, ionospheric absorption is too great and noise levels too high to give the required S/N_0 ratio; above f_2 the ionosphere cannot support the given frequency for the required percentage of time to achieve the required circuit reliability. Thus satisfactory operation can only be achieved between f_1

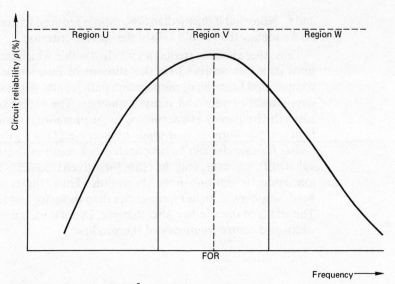

Fig. 6.7   *Circuit reliability as a function of frequency*

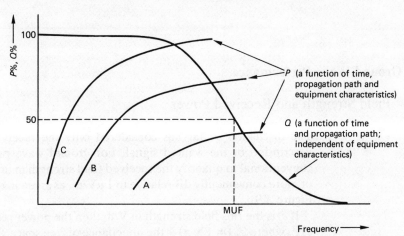

Fig. 6.8   *Dependence of P and Q on frequency and signal strength*

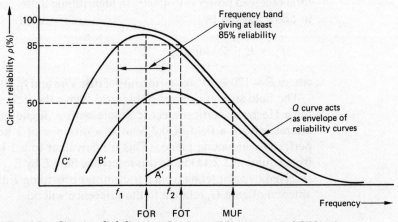

Fig. 6.9   *Circuit reliability as a function of frequency and S/N ratio*

and f_2. Note that if the median S/N_0 ratio is reduced to that for curve B', there is no frequency that would satisfy the service criterion.

Thus, there is some frequency window within which satisfactory communications could be achieved (in the absence of interference). This window is a complicated function of propagation path length, geographic location, time of day, season of year and sunspot number. The centre of this window moves along the frequency axis according to the prevailing conditions. For example, a high sunspot number year tends to increase Q for a given frequency because higher frequencies can be propagated as a result of the increase in MUF. The value of P, however, may decrease for a given frequency because of the greater ionisation, which enhances absorption. Thus, the centre of the frequency window moves to higher frequencies than those for low sunspot number years. The width of the window also changes. Diurnal variations also affect both the width and centre frequency of the window.

6.4   Ground Wave Performance

6.4.1   Field Strength and Received Power

The discussion so far has considered only the received signal power as a description of the wanted signal. For ground wave propagation it is more conventional to quantify the received field strength in units of volts per metre, or more conveniently dB relative to 1 µV/m, as given in the reference curves of Figure 3.9.

If e_r is the rms field strength in V/m then the power received per unit area is e_r^2/Z_0 where $Z_0 (= 120\pi)$ is the impedance of free space. The effective aperture area of a receiving antenna of unity gain is $\lambda^2/4\pi$, equation (6.3), and hence the total received power is $e_r^2/480\pi^2$. In logarithmic units, received signal power P_r in dBW is given by

$$P_r = E - 20 \log_{10} f_M - 107 \tag{6.15}$$

where $E = 120 + 20 \log_{10} e_r$, in units of dBµV/m and f_M is the frequency in MHz.

The field strength curves of Figure 3.9 give the field strength E in dBµV/m for a Hertzian vertical electric dipole with a dipole moment $5\lambda/2\pi$ ampere metres, giving a field of 0.3 V/m at a distance of 1 km on the surface of a perfectly conducting plane. (This is equivalent to a 1 kW isotropic radiator, from equation 3.2.) Denote this reference field E by E_R. Then the signal power received by an antenna from a transmitter providing T dBkW of power into an antenna of gain G_t relative to the reference will be

$$P_r = E_R - 20 \log_{10} f_M - 107 + T + G_t \tag{6.16}$$

6.4.2   Baseline Assessment

There are many variables which must be considered in a study of communications range capability. It is therefore convenient to adopt a 'baseline' configuration for which detailed calculations may be made. The effects of other factors either different from or not included within the baseline values can then be examined in terms of their deviations from this baseline.

The effective radiated power depends not only upon the transmitter output power but also upon the antenna efficiency and degradation factors caused by ground losses. The detailed theory of antenna radiation is extremely complicated, and it is neither practical nor desirable to consider the theory in great detail here. In the low frequency end of the HF band, antennas for mobile applications are necessarily electrically short, so that they exhibit low radiation resistance. This may have a significant impact upon communications performance. (The effect for mobile applications is described in Chapter 7.) The baseline for ground wave performance assessment is more conveniently chosen in terms of a point-to-point link, which will now be described.

Consider as baseline example a 1 kW transmitter and a vertical mast antenna of height 15 m (i.e. quarter wavelength at 5 MHz). The directivity of such an antenna in the horizontal plane compared to isotropic is approximately 3 dB; however, this will be counterbalanced by a typical 3 dB loss resulting from absorption of energy by the ground. The antenna efficiency is assumed to be -7 dB at 2 MHz, rising to -3 dB at 10 MHz. These are therefore also the values of G_t required at the given frequency in equation (6.16).

Equation (6.9) for the received signal-to-noise ratio can be rewritten as

$$S/N = P_r + D_r' - N_i \tag{6.17}$$

where P_r and N_i are the received signal and noise powers (in dBW) respectively and

$$P_r = E_p - L$$

$$N_i = kT + B + F_a$$

It is assumed that the noise levels at HF that would be experienced would most probably be akin to the CCIR[8] rural noise level. The value of F_a, the effective antenna noise factor resulting from the external noise power applied to a perfect antenna, is then

$$F_a = 67 - 27.7 \log_{10} f_M \tag{6.18}$$

The noise power level per unit bandwidth is, in logarithmic units:

$$N_0 = -204 + F_a \tag{6.19}$$

The required value for S/N_0 from equations (6.17), (6.16) and (6.10) is then:

$$S/N_0 = (E_R - 20 \log_{10} f_M - 107 + T + G_t) + D'_r + (204 - 67 + 27.7 \log_{10} f_M) \quad (6.20)$$

Figures 6.10 and 6.11 show the signal-to-noise ratios S/N_0 in a 1 Hz bandwidth (in dB) for moist ground and medium dry ground respectively for a 150 W transmitter ($T = -8$ dBkW) and a 15 m vertical mast antenna. D'_r is assumed to be 3 dB.

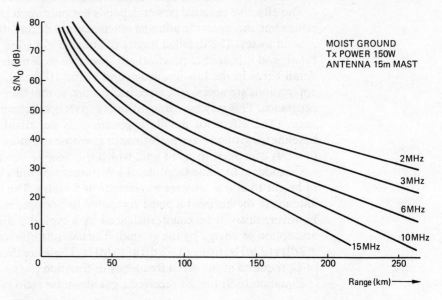

Fig. 6.10 *Signal-to-noise density ratios as a function of frequency and range over moist ground*

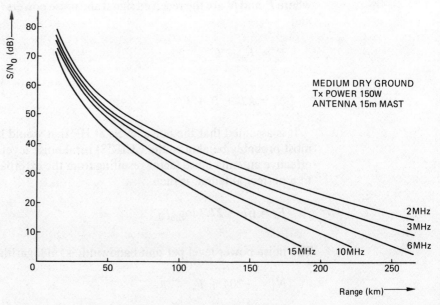

Fig. 6.11 *Signal-to-noise density ratios as a function of frequency and range over medium dry ground*

6.4.3   The Range Factor

Having established the baseline achievable ranges for a specific point-to-point link, a simple but reasonably accurate method is required for deducing the effect upon this range of changing the conditions from those given at baseline.

The propagation characteristics inherent in Figure 3.9 provide a clue. Over a range from 1 to 100 km, they almost invariably (except for 30 MHz over sea-water) provide a good approximation to a linear relationship between log (distance) and field strength (dB). Thus there is a characteristic *range factor* by which to modify the achievable range for a given change in signal-to-noise ratio. This range factor is approximately 1.12 over sea paths and 1.06 over land paths for a 1 dB change in signal-to-noise ratio.

S/N VARIATION (dB) FROM BASELINE	RANGE FACTOR TO BE APPLIED	
	Over sea	*Over land*
0	1.00	1.00
1	1.12	1.06
2	1.25	1.12
3	1.41	1.19
4	1.57	1.26
5	1.76	1.34
6	1.97	1.42
7	2.21	1.50
8	2.48	1.59
9	2.77	1.69
10	3.11	1.79
15	5.47	2.39
20	9.68	3.20
25	17.0	4.29
30	30.0	5.74

Table 6.1   Range factors for variation of signal-to-noise ratio from baseline

Table 6.1 shows the factor which should be applied to the baseline achievable range to estimate the effect of a given deviation from baseline conditions. These values are in good agreement with the theory (see Figures 3.2 and 3.3) which predicts that for frequencies in the HF band over a sea path of up to 100 km the factor should be 10 for a 20 dB change in signal-to-noise ratio. (This is characteristic of the direct radiation zone.) Over land paths for the ranges and frequencies of interest, the Sommerfeld zone is applicable up to a range of about 100 km and the factor should be 10 for a 40 dB change in signal-to-noise ratio.

The major causes of various deviations from the baseline have been identified in Chapter 3. The effect that mixed terrain type paths, irregular terrain and obstacles, scattering, vegetation, transmitting antenna characteristics and interference at the receiver can have upon the system performance has been described. These effects, either in isolation or combination, can be considered as deviations from the baseline set of conditions; they can be quantified in terms of a range factor to be applied to the baseline achievable range. Provided that the communications range of interest lies between the approximate limits of 5 km and 100 km for frequencies in the HF band, and that the comparison baseline is chosen sensibly so that large range factors are avoided, then the range factor method provides a simple, yet effective, means of quantifying the ground wave performance of any given HF communications system.

6.4.4   Effect of Short Antennas

As an example of the range factor, consider the case of a vehicle-mounted short whip antenna and its effect upon range performance compared to a baseline of 40 W isotropic radiated power. This new baseline is chosen for convenience in this example.

The impedance of an ideal electrically short monopole antenna mounted on an infinite ground plane is given by Jasik[9]. For a 2.4 m whip antenna on a large ground plane the efficiency is given approximately by

$$\eta = 2.5/[2.5 + 1.5/(kh)^3] \tag{6.21}$$

where $k = 2\pi/\lambda$
$\quad\quad h =$ monopole height

For a vehicle-mounted whip antenna a further source of loss results from absorption of energy by the ground[10]; experience with monopole antennas suggests a degradation factor in the region of 6 dB. If the transmitter output power of the vehicle set is assumed to be 40 W, then the effective radiated power is as shown in Figure 6.12.

Table 6.2 shows the deviation (calculated from Figure 6.12) from a baseline of a 40 W (16 dBW) isotropic radiator for a number of frequencies. These can be converted to range factors using Table 6.1.

To estimate achievable range, a minimum acceptable received signal-to-noise ratio must be adopted. It is proposed to use 10 dB signal-to-noise ratio in a 3 kHz bandwidth as the minimum usable criterion (as recommended by the CCIR[6]) for voice communications. Baseline conditions of a 40 W isotropic radiator over medium dry soil in the presence of rural noise result in the achievable ranges shown in Figure 6.13 (continuous curve) for the frequencies across the lower HF band. Baseline ranges for a number of frequencies are given in Table 6.2.

FREQUENCY (MHz)	BASELINE DEVIATION (dB)	RANGE FACTOR	ACHIEVABLE RANGE (km) 40 W isotropic radiated	40 W with vehicle mounted whip
2	33	6.42	67.4	10.5
4	25	4.29	55.7	13
6	20	3.20	48.0	15
8	16	2.53	41.7	16.5
10	14	2.25	39.8	17.7
12	12	2.0	38.0	19

Table 6.2   Range performance for a vehicle-mounted whip antenna with ground losses

These ranges can then be used to evaluate the effect of using a 2.4 m whip antenna with ground plane losses instead of the ideal 100% efficient antenna. The range factors in Table 6.1 can be applied to the baseline ranges to provide an estimate of the achievable ranges for the configuration of interest. Application of the range factors of Table 6.2 to the baseline curve gives the broken curve in Figure 6.13. For this communications system operating over relatively

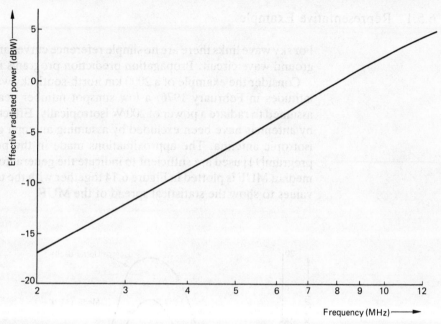

Fig. 6.12 *Effective radiated power for a 2.4 m vehicle-mounted whip antenna (40 W transmitter)*

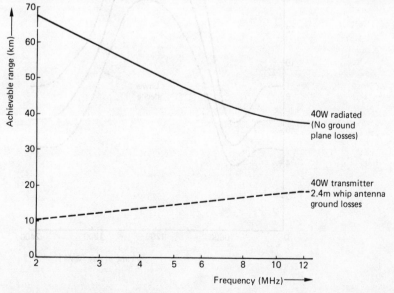

Fig. 6.13 *Achievable ground wave range for propagation over smooth medium dry soil with rural noise*

poor conductivity soil, greater ranges are achieved using the higher frequencies. The better antenna efficiencies and lower noise levels more than compensate for the extra propagation losses at the higher frequencies.

6.5   Sky Wave Performance

6.5.1   Representative Example

For sky wave links there are no simple reference curves analogous to those for a ground wave circuit. Propagation prediction programs must be used.

Consider the example of a 2000 km north-south sky wave link for European latitudes in February 1976, a low sunspot number year. The transmitter is assumed to radiate a power of 200 W isotropically. Effects such as those caused by antennas have been excluded by assuming antenna gains of unity from an isotropic antenna. The approximations made in the progagation prediction program[11] used are sufficient to indicate the general form of the results. The median MUF is plotted in Figure 6.14 together with the upper and lower decile values to show the statistical spread of the MUF.

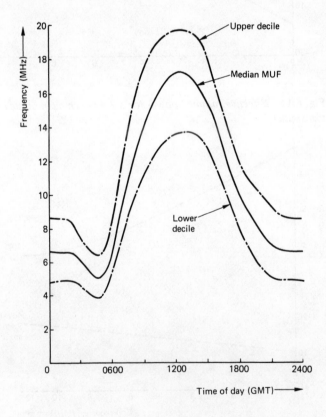

Fig. 6.14   *Predicted MUF and its variability for example radio link*

6.5.2   Monthly Signal-to-Noise Ratios

The received S/N_0 ratio in dB at the receiver output (for the days the sky wave exists) is now to be calculated.

The total noise power N_0 in a 1 Hz bandwidth is assumed to be determined by the external noise, a reasonable assumption in the HF band. The S/N_0 ratio is shown as a function of time of day and wave frequency in Figure 6.15. During the daytime the wanted signals of lower frequency are more strongly absorbed than is the noise. At night the values of the S/N_0 ratio are similar for all the frequencies. The median MUF, however, is only about 7 MHz at night (see Figure 6.14). For higher frequencies the sky wave is less strongly absorbed than at the lower frequencies but has a low reliability because Q is small. Note that by working close to the MUF throughout the day a consistently good S/N_0 ratio (broken curve in Figure 6.15) can be achieved.

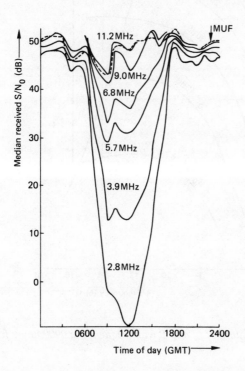

Fig. 6.15   *Predicted signal-to-noise ratios for example radio link*

6.5.3   Frequency Dependence

To illustrate the ideas behind Figure 6.8, Figure 6.16 shows, for the present example, how Q and P depend upon frequency at 0900 hours for various values of R_0, the required signal-to-noise ratio in a 1 Hz bandwidth at the receiver output. If the product of Q and P is taken then the curves of Figure 6.17 are produced. They show how ρ depends upon the frequency and required S/N_0 ratio for a given hour (cf. Figure 6.9). The greater the required S/N_0 ratio, the less the reliability. These curves can also be used to assess the improvement of circuit reliability if the equipment parameters are changed. Suppose that, for our example, R_0 is 20 dB. If the transmitting antenna is now given a 10 dB gain then the new reliability curve, appropriate to the requirement for a 20 dB signal-to-noise ratio, becomes the curve $R_0 = 10$ dB in Figure 6.17.

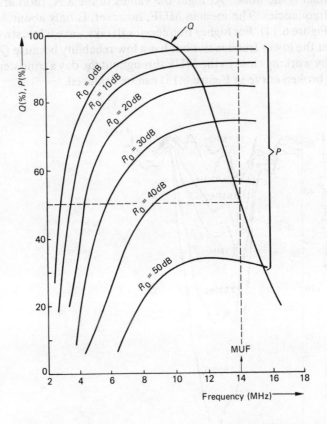

Fig. 6.16 *Variation of Q and P with frequency at 0900 hours for example radio link*

As a further illustration consider a frequency of 6.8 MHz for which (from Figure 6.15) the received S/N_0 ratio R_m is 33 dB at 0900 hours. Then Figure 6.17 shows that

if $R_0 = 40$ dB $\rho = 34\%$

if $R_0 = 20$ dB $\rho = 74\%$

Thus an increase of 20 dB in R_m (or a decrease of 20 dB in R_0) would improve the circuit reliability by 40%.

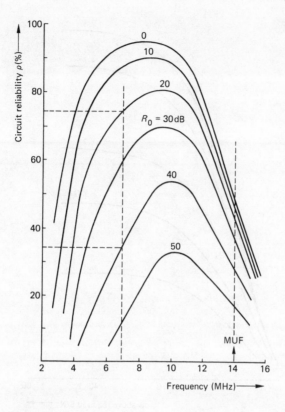

Fig. 6.17 *Variation of circuit reliability with frequency at 0900 hours for example radio link*

6.5.4 Assessment of Reliability Improvement

To assess adequately the performance of a sky wave link, the communications engineer needs to know what improvement in circuit reliability can be achieved by a given increase in the received S/N_0 ratio. The increased value of ρ does not depend on how the improvement is made; for example it might be achieved by an increase in transmitter power, use of directional receiving antennas, improved transmitter antenna efficiency, etc.

Figure 6.18 shows the values of reliability calculated for the point-to-point link example discussed above. It is convenient to label the abscissa as 'dB above required signal-to-noise' and thus a direct comparison of the effect of different frequencies is made, although different initial transmitter powers or system gains are required to produce the same signal-to-noise value for different frequencies.

Thus, for example, if the signal-to-noise ratio on the link is improved by 10 dB by some means (increased transmitter power, improved efficiencies, antenna gain, etc) the corresponding improvement in reliability can be estimated[12] from the curves. The following points should be noted:

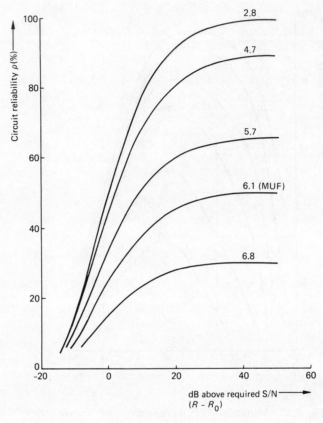

Fig. 6.18 *Reliability improvement for different frequencies at 0300 hours for example radio link*

1   For large values of $(R - R_0)$, ρ tends to the value Q since P tends to unity. This means that well below the MUF $\rho \approx 1$, while as the frequency is increased ρ decreases for a given signal-to-noise ratio. At the MUF, $\rho = \frac{1}{2}$ and above it $\rho < \frac{1}{2}$.

2   For very large negative values of $(R - R_0)$, $\rho \approx 0$ for all frequencies, since the probability that the instantaneous S/N ratio exceeds the required level R_0 is very small.

3   Between the extremes **1** and **2** the curves are approximately linear, passing through the value $\frac{1}{2}Q$ at $R = R_0$. The slope is determined by the statistical spread of absorption loss and external noise characteristics.

4   The slopes of the curves in Figure 6.18 are proportionally reduced by the value of Q. Thus the MUF ($Q = \frac{1}{2}$) has a slope of one half of that for frequencies which have $Q = 1$.

5   Figure 6.18 appears to show that the lower frequencies are the most reliable. This is only true if the actual received signal-to-noise ratio for each frequency is the same. In this case the lower frequencies are advantageous because of the more favourable value of Q. However, the lower frequencies are often more heavily absorbed, particularly during the daytime.

7 Air-Ground Communications

7.1 The Mobile User

7.1.1 Comparison with Point-to-Point Circuits

High frequency circuits that propagate via the ionosphere are used extensively for long-range point-to-point communications and broadcasting. Characteristics of such links have consequently been studied extensively; commercial services are available for prediction of optimum working frequencies and the quality of communications at these frequencies. Most point-to-point land-fixed HF communications circuits use high-gain rhombic or log-periodic antennas, whilst arrays of horizontal dipoles, also with significant directivity, are popular for broadcasting using the sky wave mode. In principle, therefore, the capability of these links may be optimised by good engineering design and practice in respect of the equipment and antenna systems, whilst high transmitter power is often available.

Much more difficult problems, however, are presented by HF communications to mobiles. Communications are often required at ranges from a few kilometres to several thousand kilometres over a wide variety of terrain; this implies different modes of propagation according to range. Physical constraints are placed upon the antenna so that its efficiency may be degraded; radiation patterns are obtained that may not be suited to the propagation mode. The primary transmitter power may be constrained severely for the mobile user, whilst serious excess noise, both acoustic and electrical, may be present at the mobile terminal. Additionally for an aircraft, its height may give rise to further multipath propagation mechanisms, whilst its speed may cause Doppler frequency shifts.

To achieve satisfactory results over an HF link of this kind, careful consideration must be given to the terminal radio equipment, the planning of operational links and the management of the frequencies to be used over those links. An understanding of the overall system considerations is essential to the satisfactory design and operation of HF mobile radio links.

7.1.2 Parameters Critical to the Mobile User

From equation (6.9) it was shown that the received signal-to-noise is given, in logarithmic units, by

$$S/N = E_p - L + D'_r - N_i \tag{7.1}$$

where E_p is the effective radiated power
$\quad\;\;L$ is the propagation path loss
$\quad\;\;D_r'$ is the receiving antenna directivity factor against far field noise
$\quad\;\;N_i$ is the noise power incident at the antenna.

The comments made above, in section 7.1.1, can be related to the components of equation (7.1) in terms of three attributes of the mobile terminal:

1   Physical Constraints

The physical size and shape of the mobile may limit E_p in two respects. The primary transmitter power may well be much less than for a static installation whilst the constraints upon size and siting of a suitable antenna may severely limit both its efficiency and polar diagram. The net effect will be that E_p may be many decibels below that for a typical fixed ground station. The parameter D_r' is also affected by the limitations placed upon the mobile's antenna. It may be impossible to provide any directional properties for the antenna. Thus discrimination against far field noise, often provided at a fixed site, is extremely difficult to achieve for the mobile.

2   Environmental Constraints

The local interference environment of a mobile user is likely to be much more severe that for a fixed terminal, thus N_i is much larger. Many electrical equipments may be housed within a relatively small volume, causing high levels of noise to the HF receiving antenna. In contrast, at fixed ground stations the receiving antennas can often be sited at a considerable distance from potential interfering equipments.

3   Mobility Constraints

Propagation path loss L can be minimised by selection of an appropriate frequency but this choice depends upon many factors including communications range. Large changes in communications range over relatively short time intervals, as would be experienced by rapidly moving mobiles particularly aircraft, imply the need to change frequency very regularly. Because of the need to continually optimise the communications frequency it is very likely that the path loss L for the mobile user will be greater than that for an equivalent point-to-point circuit.

The overall effects of the constraints imposed upon the mobile user are that E_p and D_r' will be less than, and L and N_i will be greater than, for a corresponding point-to-point link. Thus from equation (7.1) the signal-to-noise ratio may be reduced dramatically for the mobile user. The problems that need to be addressed are many and varied. It is convenient to consider a particular example to highlight the fundamental principles involved. An excellent illustration of all of the above limitations is provided by a study of air-ground communications, the subject of the present Chapter.

In order to understand the impact upon communications reliability and performance it is necessary to examine first the limitations imposed by the aircraft terminal itself. These limitations, which stem from **1, 2** and **3** listed above are now to be discussed in terms of: Antenna related effects; Noise environment; Frequency selection.

7.2   Characteristics of the Airborne Terminal

7.2.1   Antenna Radiation Efficiency

HF antenna systems on aircraft comprise the radiating structure, ancilliary devices such as RF switches and reactive loading components, the antenna tuning and matching unit (ATU) and the feeder cable. The importance of antenna siting is crucial.

The airframe tends to be the dominant radiator at frequencies near its natural electrical resonances. For example, a half-wave longitudinal resonance occurs in a small fixed-wing aircraft at about 9 MHz; if the aircraft is fitted with a notch antenna the airframe radiation is dominant between about 5 to 15 MHz. It so happens that this frequency band is also suitable for long-range sky wave communications, for which the aircraft may be regarded as a horizontal half-wave dipole.

Although the radiation efficiency η of an aircraft antenna may be very small in some circumstances, receiver noise at the aircraft terminal is generally dominated by the external noise field. Thus, provided that radiation efficiency exceeds a threshold value, further increase in radiation efficiency provides no advantage for reception only and the relevant antenna parameter is its directivity D. This is obtained as a function of angular coordinates and polarisation from the normalised radiation patterns.

At the low end of the HF band poor radiation efficiencies[1] are exhibited by small and medium sized aircraft for the following reasons:

a) Many aircraft HF antennas are electrically extremely small.
b) The whole airframe (regarded as a radiator) is itself electrically small.
c) There are constraints upon antenna type and siting.

A very adverse combination of factors occurs in small, fixed-wing, high-performance aircraft, for which radiation efficiencies approaching –50 dB at 2 MHz have been measured. Thus for 100 W transmitter power only some 10 mW is radiated, the rest being dissipated in loss resistance. Typical antenna radiation efficiencies for small, medium and large aircraft are shown in Figure 7.1.

At most frequencies the antenna exhibits a highly reactive impedance so that an ATU is required to tune it to the operating frequency and to present a good impedance match to the transmitter. For transmission bandwidth calculations the antenna system may be treated as a tuned circuit and sometimes very high *circuit quality factors* (Q values) are encountered implying very small transmission bandwidths. Usually the bandwidth of a tuned circuit is defined at the –3 dB points, giving

$$B = f_0/Q \tag{7.2}$$

However, in an aircraft HF radio system this definition is inappropriate for both transmission and reception. On transmission there is an engineering limitation on the maximum VSWR which can be tolerated by the power amplifier stage. The –3 dB point of a tuned circuit corresponds to a VSWR of

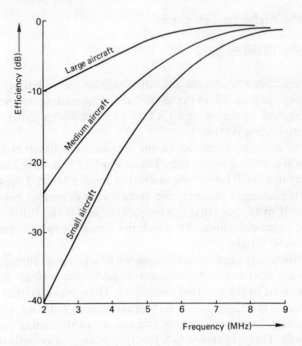

Fig. 7.1 *Frequency dependence of aircraft antenna efficiencies*

5.8:1, which is much higher than current transmitter specifications allow. Since the receiving system performance should be externally noise limited, the relevant antenna system criterion on receive is that the radiation efficiency (taking account of impedance mis-match) should exceed a threshold value determined by the values of the external noise field and the receiver noise figure. Generally the useful antenna bandwidth determined in this way is much greater than the –3 dB bandwidth.

7.2.2   Antenna Radiation Patterns

As the radio frequency is increased from 2 MHz, radiation efficiency increases, whilst radiation patterns may vary considerably. These characteristics depend on the antenna/airframe combination.

Ideally the radiation patterns should be matched to the propagation mode that needs to be used. It is convenient in the first instance to treat the aircraft antenna as an elevated electric dipole as follows:

a) Ground wave. A vertical dipole gives the required vertical polarisation and omni-azimuth coverage.
b) High angle sky wave. A horizontal dipole gives the required high angle coverage, azimuth orientation being immaterial.
c) Low angle sky wave. Either a vertical or a horizontal dipole may be employed, but the latter exhibits reduced directivity for angles near the dipole axis.

Operation in the low frequency end of the HF band is necessary for both ground wave and high angle sky wave links, so that from *a* and *b* above there are conflicting antenna requirements. Long-range sky wave links employ higher operating frequencies and such conflicts do not necessarily occur in this case.

Some types of aircraft may be fitted with tail-fin notch or vertical loop antennas which radiate as magnetic dipoles. Such antennas give good high angle coverage together with vertical polarisation in the azimuth plane, but the radiation patterns exhibit broadside nulls. Thus they offer reasonable compromise radiation patterns for both ground wave and high angle sky wave links. Aircraft-installed antennas are compound radiators and in some cases it is possible to configure the antenna to give vertically polarised omni-azimuth coverage whilst at the same time the airframe gives high angle coverage.

Few radiation patterns of large aircraft have been measured, but at frequencies when their length is much greater than half a wavelength they exhibit structured radiation patterns, giving degraded link performance near directions of the minima. As a simple working rule the number of lobes (and minima) at wavelength λ for a cylindrical dipole of length L is given by

$$n = 4L/\lambda \qquad (7.3)$$

The lobes exhibit cylindrical symmetry. For a cylinder of length 50 m (roughly representing a large aircraft), a 4-lobe radiation pattern is exhibited at 6 MHz, with nulls broadside to the axis. At frequencies of 9, 12 and 15 MHz, 6-, 8- and 10-lobe radiation patterns respectively are exhibited. Due to the non-cylindrical shape of the aircraft the radiation patterns exhibit departures from the cylinder case, but remain highly structured with consequent communications link reliability implications.

7.2.3 Aircraft-generated Noise

Probably the most important feature of the ground-to-air link is the noise environment of the aircraft. All aircraft systems which use electrical energy are likely to generate unwanted electromagnetic energy, and this may couple into the aircraft radio systems and degrade their performance. Conversely, almost every aircraft radio transmitter generates intense electromagnetic fields which may effect other avionic systems, including installed radio systems operating at the same time. Coupling mechanisms between an interference source and the rest of the avionic installation may be complicated; interference levels are affected by factors such as design, practice and workmanship of the avionic installation, imperfect shielding of braided coaxial cables and the RF attenuation offered by the aircraft skin. A detailed discussion of these effects has already been given in section 5.3. From the estimates of aircraft-generated noise given in section 5.3.5 it can be seen that aircraft noise levels in the HF band may be 15 dB or more above those at a ground station in a rural location.

7.2.4 Flight Paths and Frequency Selection

In considering the use and performances of HF communications systems for aeronautical purposes it is convenient to categorise the usage into en-route and off-route applications. Civil aircraft flying the North Atlantic route between Europe and North America provide a good example of en-route usage. The routes are laid down geographically and most flights are scheduled. The aircraft need to communicate over long ranges by HF. This is undertaken in order to give positional information at certain times and such information as estimated time-of-arrival, fuel state, airfield diversion, etc. These messages from the aircraft are relatively short and usually require a short acknowledgement from the ground. If the information does not get through first time but takes a few minutes or so, this is usually acceptable.

The communications frequencies to be used, both primary and secondary, are provided at crew briefing. Because the tracks flown are well charted, past experience helps a great deal in achieving reliable communications. Communications are generally relatively satisfactory, but of course problems occur during ionospheric disturbances. For long-haul transport routes in temperate and lower latitudes, high availability and reliability is usually achieved.

A different picture emerges, however, for off-route usage. Military aircraft fly on routes that are not regularly used. They include large aircraft with crews which may or may not include a specialist radio operator, and high performance aircraft with only a one or two man crew who use the radio as just one more item of airborne equipment. In these cases it could be that the aircraft does not wish to transmit unless it is essential, but that when it does, the response of the communications system must be fast and highly reliable. The choice of the correct frequency is in this case important. As an example take an aircraft operating to the north of the UK in summer 1977 and wishing to communicate with a base in southern England. Table 7.1 indicates the frequency[2] that should be used depending on time of day and distance from the base. Up to eight frequencies could easily be required for a typical sortie.

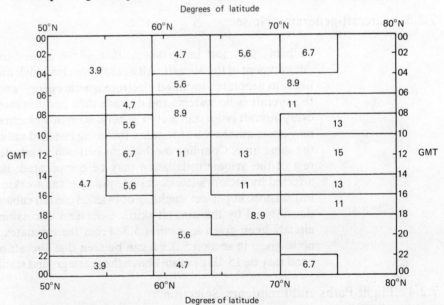

Table 7.1 Optimum frequencies (in MHz) for aircraft communication to a site in southern England in summer 1977

7.3   Impact upon System Performance

7.3.1   Effect of Antenna Efficiency

Consider now how the properties of the aircraft antenna would change the shape of the reliability curve already given in Figure 6.7. A typical graph of aircraft antenna efficiency against frequency[3] is shown schematically in Figure 7.2a. The frequencies f_{10}, f_{20} ... designate antenna efficiencies of -10 dB, -20 dB...and are different for large and small aircraft as already seen from Figure 7.1. The modified reliability curve for $R_0 = 20$ dB in Figure 7.2b is then given by the chain curve, which passes through the $R_0 = 30$ dB curve at $f = f_{10}$, $R_0 = 40$ dB curve at $f = f_{20}$ and so on. The effect of the frequency variation of antenna efficiency is therefore to shift the original $R_0 = 20$ dB curve towards the higher frequencies and this restricts the frequency range capable of providing a given reliability ρ' from $(f_2 - f_1)$ to $(f_2 - f_1')$, see Figure 7.2b. The width of the frequency window is reduced. The centre of this window moves along the frequency axis according to the prevailing conditions as discussed in section 6.3.5.

Figure 7.3 shows a schematic example to illustrate these effects. The result of the poorer antenna efficiency on the small aircraft is twofold relative to the large aircraft:

a) The usable frequency range is decreased; from 4.4 – 11.8 MHz to 6.3 – 11.8 MHz for $\rho = 60\%$.

b) The reliability at the lower frequencies is decreased, from 60% to 20% at 4.4 MHz.

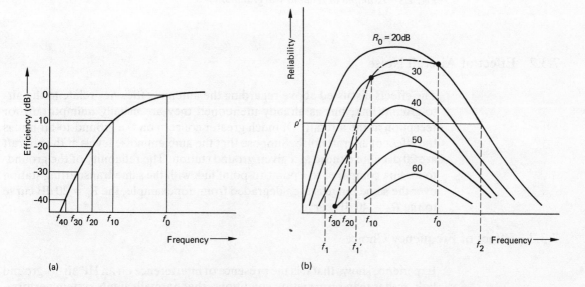

(a)   (b)

Fig. 7.2   *Effects of degraded aircraft antenna efficiency on circuit reliability*

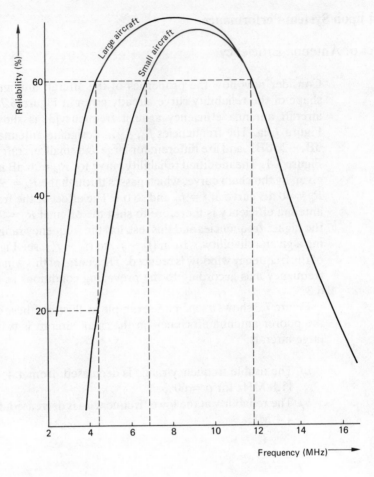

Fig. 7.3 *Example of reliability degradation*

7.3.2 Effect of Aircraft Noise

The effects discussed above regarding the antenna efficiency relate to the air-to-ground link, but as already mentioned they are usually unimportant for reception at the aircraft. Of much greater concern on the ground-to-air link is the effect of aircraft noise. Suppose that the ambient noise levels at the aircraft are 20 dB above those at a given ground station. The reliability of the ground-to-air link compared to a point-to-point link with the same transmitting station over the same range is then degraded from, for example, the $R_0 = 20$ dB curve to the $R_0 = 40$ dB curve.

7.3.3 Effect of Frequency Choice

Experience shows that it is the presence of interference on an HF air-to-ground link, rather than propagation conditions, that normally limits system performance. The interference is found to be least during the daytime at frequencies

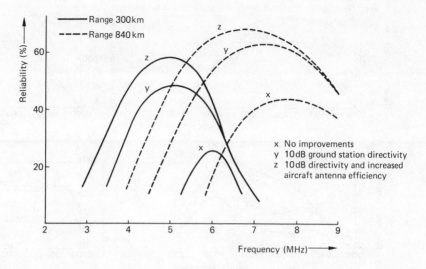

Fig. 7.4　*Variation of reliability of an air-ground voice link*

well below the MUF; it may sometimes be advantageous to operate below the optimum working frequency accepting some loss in propagation quality in exchange for less interference from other spectrum users.

The dependence of reliability upon frequency is shown in Figure 7.4, which gives predicted values[4] for a small aircraft operating over two short-range links at noon in January 1976. Note that at the lower frequencies there is a rapid decrease in reliability due to the decrease in antenna efficiencies (see Figure 7.1). The curves X to Z show examples of progressive improvements which might be made. Curve Y shows the effect of a 10 dB ground directivity factor whilst curve Z shows how, in addition, increasing the aircraft antenna efficiency could increase reliability. January 1976 was a period of low sunspot activity, implying that operating frequencies were well down in the region where aircraft antennas are least efficient.

Progressive improvements in link reliability may be achieved by improving aircraft antenna efficiency, ground antenna directivity and by working over longer-range links. Ground antenna directivity gives improved link performance when the system is limited by external noise, but is of little help where reliability of the propagation path is itself a major constraint. The curves in Figure 7.4 demonstrate the need for good frequency management, but even with some favourable assumptions about the communications terminals the link reliability often falls short of what would be regarded as desirable, particularly at night. Propagation modes can fail very rapidly, so that if frequency management is poor the effects on communications can be very serious.

7.3.4　Effect of Flight Path

Conditions are likely to change as a result of aircraft flight path and these changes can affect the overall reliability for a typical mission profile[4].

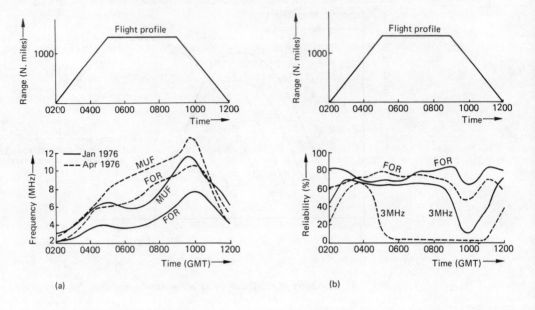

Fig. 7.5   *Variation of frequency and reliability for a typical aircraft flight at dawn*

Consider an example of air-ground voice communications utilising 10 dB ground receiving antenna directivity and assume that the mission is of 10 hours duration. The ground station is in northern Britain and the aircraft flies northward at 450 knots for 3 hours, remains at this range for 4 hours and then returns, again at 450 knots, during the last 3 hours of the mission. The period is one of low sunspot activity.

In Figure 7.5a predictions are shown for the above mission during January and April 1976, a low sunspot number year. Curves are given for the MUF and for the frequency of optimum reliability (FOR), given by the frequency which provides the best reliability (this is not necessarily the optimum working frequency). Figure 7.5b shows the reliability factor for this best frequency (FOR) and the reliability factor for a given fixed frequency (3 MHz). The curves take account of a 10 dB ground directivity assuming that it was maintained throughout the flight.

Figure 7.5 shows a mission starting at 0200 hours. The optimum frequency is a compromise between sky wave availability and signal absorption. Note that the optimum reliability (FOR curves in Figure 7.5b) attainable is approximately constant over the whole mission. For a constant (3 MHz) frequency the January results (continuous curves) show that this would be a reasonable frequency for most of the mission, but when the MUF increases (around 1000 hours) the reliability of 3 MHz decreases and the FOR rises to about 8 MHz. Results for April (broken curves) show that the FOR is greater than for January and hence 3 MHz is not a good frequency choice since the MUF is well above 8 MHz for most of the mission. In this case 3 MHz signals are too heavily attenuated to be useful.

7.4   Communications with Small Aircraft

7.4.1   Summary of System Problems

It is desirable that communications performance on the ground-to-air and air-to-ground paths should be similar. Compared with a good ground station, the aircraft terminal is degraded both on transmission and reception; separate assessments of link reliability in both ground-to-air and air-to-ground directions are therefore necessary.

In order to examine these factors in more detail the example of a small aircraft will be taken since this provides an excellent illustration of key factors contributing to the degradation of the communications systems performance. It is common experience[5] that at the shorter ranges (50–500 km) HF communications can be unreliable to small aircraft. These difficulties arise as a result of one or more of the following:

a) For small aircraft the antenna efficiencies are poor at the low end of the HF band. This fact coupled with the limited available transmitter power means that the effective radiated power is of the order of a few watts or less at these frequencies.

b) For these shorter ranges, the sky wave mode with frequencies in the range 2–6 MHz must often be used. It is at these frequencies that the small aircraft has poor antenna efficiencies. The problem is particularly acute at night when frequencies at the low end of the 2–6 MHz range must be used and where aircraft antenna efficiency is very poor.

c) The ground station antenna system often has inadequate high angle (>45°) coverage. Moreover the range of sky wave angles of elevation to be covered, together with the wide azimuthal coverage required, inhibits the use of good directive antennas on the ground.

d) The rapid variation of optimum working frequency at short ranges imposes an aircrew workload which is, at best, unacceptable and at worst, unachievable in a high performance aircraft.

e) Interference is high at night in the 2–6 MHz band as propagation conditions restrict the use of higher frequencies and users crowd into the low end of the HF band.

f) Electrical and acoustic noise generated by the aircraft can be very troublesome, particularly when the aircraft operates at low altitude.

Faced with problems of this complexity, the communications system analyst seeks to isolate a number of problem areas for study.

7.4.2   Performance Predictions

The predicted median signal-to-noise ratios for three sky wave circuits are given as examples, assuming that atmospheric noise is the predominant noise source at the receiver site. Three ground stations A, B, and C are used in the computations[5] at successively increasing ranges from the aircraft. The density of shading in Figures 7.6 to 7.8 shows increasing values of received signal-to-noise density (S/N_0) for a given frequency at a specified hour of the day.

Because S/N_0 is a median value, the achievement of a given criterion gives a communications reliability of approximately 50%. Thus diagonal lines in these examples indicate S/N_0 values of between 45 dB and 55 dB. They show that 50% reliability or greater would be produced for a 10 dB signal-to-noise criterion. This is equivalent to SSB voice in a 3 Hz bandwidth with $S/N_0 = 45$ dB. Where no shading is shown in the diagrams S/N_0 is less than 45 dB; SSB voice transmissions would at best be barely acceptable under those conditions and communications reliability would be unacceptable.

The highest available frequency for adequate communications at a given hour is taken to be the FOT (optimum traffic frequency) defined here as the sky wave frequency available for 85% of the time. The lowest usable frequency (LUF) depends upon the required S/N_0 criterion, and thus an improvement in the level of received signal or a reduction in the noise level will lower the LUF. The FOT, however, is independent of S/N_0 criteria, being only a function of the prevailing ionospheric conditions, which in turn depend upon hour of the day, season of the year and sunspot number.

7.4.3   Air-to-Ground Links

Figure 7.6 shows the predicted S/N_0 ratios for communications with ground station A during 1976, a year of low sunspot activity. Communications are more difficult in summer months than in winter, partly as a result of increased atmospheric noise levels at the ground station. Between 1800 hours and 0600 hours (i.e mainly dusk-night-dawn) the S/N_0 ratio does not reach 55 dB at any time of year.

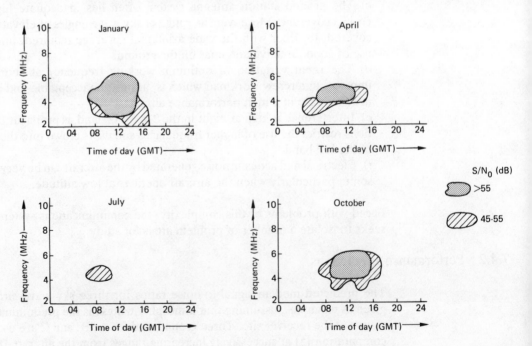

Fig. 7.6 *Short-range communications (300 km) from aircraft to ground station A during 1976*

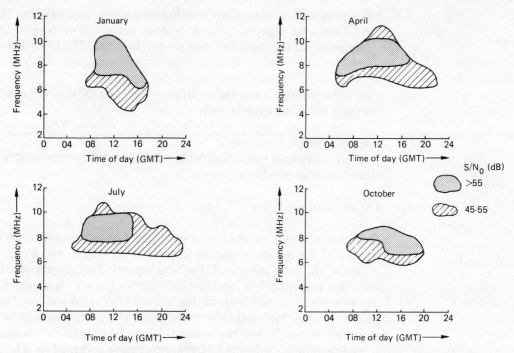

Fig. 7.7 *Medium-range communications (1200 km) from aircraft to ground station B during 1976*

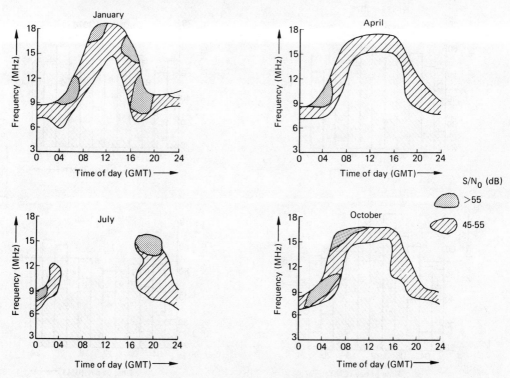

Fig. 7.8 *Long-range communications (2500 km) from aircraft to ground station C during 1976*

Consider now, ranges of communication greater than 1000 km. Figures 7.7 and 7.8 show the expected S/N_0 ratios at receiving stations B and C for aircraft transmissions in the same low sunspot number year. The following should be noted:

a) Between 2000 hours and 0400 hours the S/N_0 ratio does not reach 55 dB at station B for any time of year.

b) For the long-range link (to station C), the received S/N_0 ratio at the FOT tends to be greater at night.

c) Over the long-range link S/N_0 values for the daytime tend to be poor compared to the shorter ranges.

7.4.4   Advantage of Multi-Ground Station Usage

The value of working over a longer-range link is shown schematically in Figures 7.9 to 7.11 for the above ground stations A, B, and C, and for D, a ground station at the same range as C but well separated geographically. Figure 7.9 shows the maximum S/N_0 obtainable for any frequency for a given hour of the day by a small aircraft transmitting in both high (100) and low (15) sunspot number years of 1980 and 1976 respectively. The shaded areas show reception at ground station A and the letters B, C, and D designate the more remote receiving stations. A 55 dB S/N_0 criterion can be achieved for a large percentage of the day (see Figure 7.10) by using these longer-range links. If ground

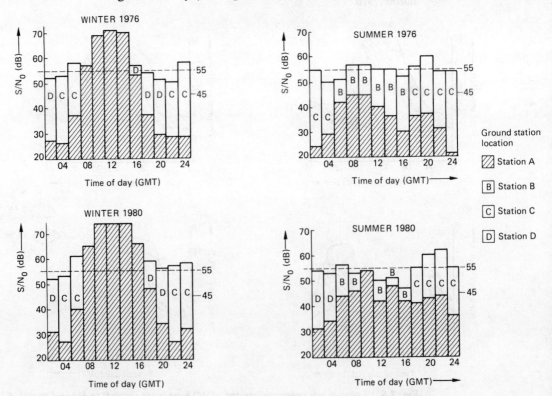

Fig. 7.9 *Optimum achievable signal-to-noise ratios for different ground stations*

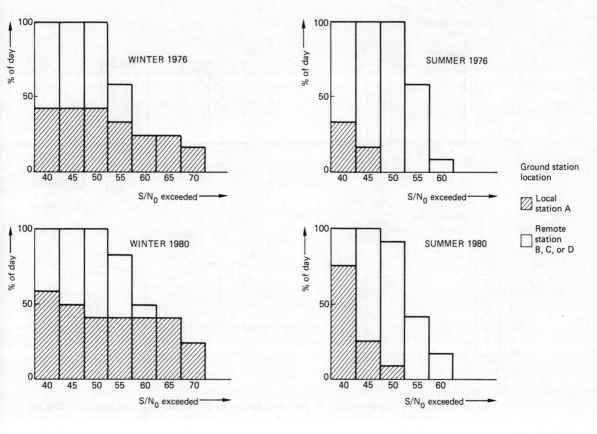

Fig. 7.10 *Percentage of the day that a given S/N_0 ratio is exceeded for air-to-ground link*

station A only were to be used for reception, a 55 dB criterion would be achieved for only a small percentage of the day in winter and not at all in the summer months.

A particularly valuable feature of using more than one remote station is that a fairly constant (to within 10 dB) S/N_0 can be achieved throughout a 24 hour period provided that different ground stations are utilised at different times of day. Moreover, if the remote stations are at various ranges and on different bearings from the aircraft, careful planning could virtually eliminate the need for frequency changing.

Figure 7.11 demonstrates the consistency of frequency for a given range and period of the day. In the winter months the frequency used is 7.5 MHz and in summer 9 or 10.5 MHz depending upon solar activity. The medium-range link (1200 km) to station B occupies the daylight hours. The longer ranges to stations C and D cover the difficult dawn-dusk period and night-time conditions between them, as a result of their geographical separation. This example of a constant frequency can be compared with Figure 7.9 which shows the effects of using the optimum frequencies. Only a few dB are lost by careful choice of the constant frequency and using all three remote stations. This

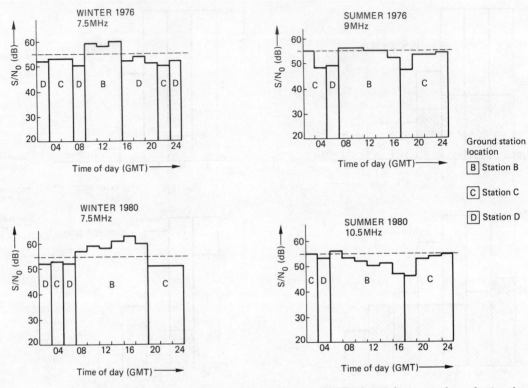

Fig. 7.11 *Communications performance using a constant frequency but selecting the best ground station*

situation would help to relieve aircrew workload of constantly changing frequencies. Advantages of the longer range links are summarised in Table 7.2.

7.4.5   Ground-to-Air Links

Since the poor aircraft antenna efficiencies are relatively unimportant in the receive mode it is ionospheric absorption that is the most important frequency dependent parameter in the link. This, in turn, implies that frequency management is relatively more important than for the air-ground situation. It is assumed that the transmitting antenna efficiencies for the ground station are not strongly frequency dependent.

For the levels of noise in the aircraft assumed here, similar calculations to those performed for the air-ground link show that a 1 kW effective radiated power from the ground transmitter is not adequate; 10 kW (40 dBW) would be more suitable. Communications would still be difficult during the daytime, however, particularly during the summer months. Under these circumstances 50 dBW would be more appropriate. For the shorter-range links the available frequency band is severely limited and there may be considerable problems with interfering stations, particularly at the lower end of the HF band. For the

REQUIREMENTS FOR REMOTE STATION WORKING	CONSEQUENCES OF USING REMOTE STATIONS
Only frequencies above ~7 MHz are required.	1 Poor aircraft antenna efficiencies are avoided. 2 Interference is reduced at night since lower frequencies are employed by other users. 3 Ground antenna size is reduced.
Antennas require only narrow azimuthal beam-width and low angle coverage	1 Good directivity should be obtainable; thus noise and interference can be discriminated against. 2 Antenna steering techniques are not necessary.
Aircraft should be at least 1000 km from remote stations	1 Frequency changing with time of day is not necessary as often as for shorter-range links. 2 Changes in aircraft position will not necessitate frequency changing 3 For aircraft in northern waters reflection of signals in auroral regions can be avoided.
Two or three remote stations ideally required, with adequate geo-graphical separation.	1 Whole 24 hour period can be covered by using ranges of ~1200 km by day, ~2500 km by night. 2 Frequency changing can be effectively eliminated by careful choice of ground station operation. 3 Pilot workload is reduced by using a single frequency.

Table 7.2 Factors connected with using remote ground receiving stations

longer-range links the available frequency band is wider but the ionospheric absorption tends to be stronger for a given frequency due to the increased propagation path length.

Some of the results for the ground-air link are summarised in Figure 7.12 which shows the effect of increasing the radiated power from the transmitter at station B. It is clear that 1 kW (30 dBW) does not provide adequate communications reliability since at no time does the reliability figure reach even 50%. An appropriate 'operationally accepable' figure would be closer to 80%. Under these circumstances 50 dBW of radiated power is generally required. This implies some directional gain on the part of the ground transmitting antenna.

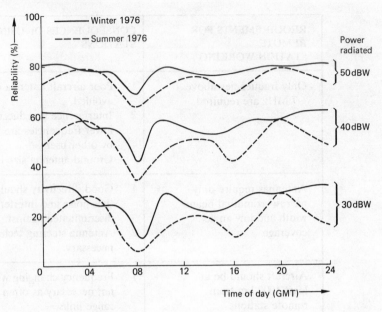

Fig. 7.12 *Ground-to-air communications reliability over a 1200 km link and its dependence upon radiated power*

7.5 Guidance for the System Designer

7.5.1 Criteria

Some guidance can be offered from Chapter 6 for the design and operation of air-ground HF communication links. It is first necessary to choose some reliability factor criteria, bearing in mind that these may be much lower for mobile communications than for commercial point-to-point links. For the present purposes $\rho = 20\%$ (nearly useless) and $\rho = 80\%$ (satisfactory) have been chosen.

In common with fixed point-to-point links it is most important to choose an operating frequency appropriate to the propagation and noise conditions prevailing at the time. Provided that this is done, the margin between nearly useless and satisfactory communications can range from 20 dB to 35 dB depending upon the time of day. Frequency allocations and interference may limit the choice of frequency.

Communications in a variety of circumstances are commonly required; both ground wave (for short distances) and high angle sky wave (for medium distances) propagation paths need also to be considered if an effective overall assessment of communications reliability is to be made. The value of R_0 needs to be established for particular cases and this may be influenced by several factors associated with the aircraft.

The reliability of an HF air-ground sky wave link is a complicated function of a large number of factors. Recourse has therefore been made to representative examples to illustrate the main phenomena. Based on these examples the objective of this section is to provide guidance for the design and management of such links.

The three most important parameters have been shown to be frequency management, antennas and aircraft noise. All these features are important when assessing an HF air-ground link since the power budget is usually more critical than for a fixed point-to-point link. Suppose that a 10 dB improvement in median received signal-to-noise ratio would provide a worthwhile increase in communications reliability for a given air-ground link. It is then necessary to decide upon the most cost-effective solution. It might be better, for example, to improve the ground station directivity factor by 10 dB rather than increasing the aircraft transmitter power tenfold or improving the antenna efficiencies of a large number of aircraft.

7.5.2 Frequency Management

Consider firstly the link management aspects. The most crucial factor is the choice of operating frequency for a particular set of conditions. The upper frequency bound is determined by the availability of a suitable propagation path, which is governed primarily by the MUF. The lower frequency bound is determined by the absorption within the ionosphere, so that it is a function of the performance of the communications terminals. The useful frequency band available to the communicator depends not only upon the signal strength and fading characteristics, but also upon the dispersion caused by multipath propagation and ionospheric movements. The normal prediction methods should therefore ideally be supplemented by daily forecasts and perhaps ionospheric soundings, as discussed in Chapter 8. The system designer needs predictions of frequencies which have tolerable dispersion in addition to adequate signal-to-noise ratios. If a choice of frequency is available, the highest frequency which offers acceptable reliability is often to be preferred since this minimises multipath effects (i.e. f_2 is preferable to f_1 in Figure 6.9).

Ground-to-air and air-to-ground communications must be treated as separate situations in view of the different transmitter powers, radio frequency noise field and antenna characteristics at the ground and airborne terminals. The MUF increases with range and if a choice of ground stations is available it may sometimes[6] be advantageous to work to the more remote station so that higher working frequencies can be used. Consequently the need for continual frequency changing would be eliminated; the sky wave propagation modes would also be operable over longer periods of a flight compared to those for a shorter range link.

7.5.3 Antennas

Particularly severe problems are presented by HF communications with small aircraft at moderate ranges, primarily as a result of very poor aircraft antenna efficiency in the low part of the HF band. Working over a longer-range communications link increases the optimum traffic frequency, thus avoiding the poor antenna efficiencies and generally reducing the external noise levels.

If all-round coverage is not required by the ground receiving station, then an important parameter is the directivity (i.e. a geometrical factor) rather than absolute gain of the ground antenna. If the noise arrives uniformly from all directions the improvement in signal-to-noise ratio is directly proportional to the receiving antenna directivity. In practice considerable anisotropy can occur in the angular distribution of noise; in this case it is still valid to include a term D'_r in equation (7.1) but it must now be regarded as a discrimination factor (against noise) rather than as a true geometrical directivity. Ground antenna directivity gives improved circuit performance when the system is limited by external noise, but is of little help when availability of the propagation path itself is a major constraint.

7.5.4   Aircraft Noise

Sources of aircraft noise are important factors for consideration and are often the predominant noise sources received in the aircraft. They include electrostatic discharges, rotating electrical machinery and switching transients so that the noise is likely to contain both Gaussian and impulsive components. Radiated interference from the avionic installation is picked up by the aircraft antennas; it is important to provide adequate attenuation of this noise by the aircraft airframe. This is a particularly difficult problem if the hull is made of non-metallic materials, as in the case of some helicopters. In unfavourable circumstances and if no steps are taken to reduce it, the intensity of precipitation static noise can exceed all other noise sources in the HF band. It is therefore necessary to have adequate static dischargers available. The amplitude distribution of the locally generated aircraft electrical noise may well be considerably different from that of other noise sources. This factor may be important for some modulation systems.

8 Frequency Management

8.1 Techniques

Satisfactory HF communications cannot be established without an ability to operate over the radio link using an appropriate frequency; the choice of the right frequency at the right time is fundamental to maintaining an acceptable communications performance. The process of selecting the best frequency according to the prevailing conditions is known as *frequency management*.

Successful frequency management depends upon the ability to predict, measure and react to a range of parameters that characterise both the propagation path and the noise. There are three broad categories of technique associated with frequency management:

1 *Predictions* These can be further subdivided into
a) Long term – required for communications systems planning, engineering design and channel allocation. Long-term predictions must cover a complete sunspot cycle.
b) Short term – required for day-to-day circuit operation, daily frequency planning and operator briefings.
2 *Soundings* These provide an up-to-date indication of propagation characteristics over vertical or oblique paths. The likely impact of the prevailing conditions upon wanted signal parameters can be deduced in a semi-empirical manner.
3 *Channel Evaluations* These provide real time channel quality assessments taking account of prevailing propagation, noise and interference effects.

The above three techniques are listed in increasing order of complexity and effectiveness. They are discussed in more detail in the following sections.

8.2 Ionospheric Predictions

8.2.1 Predicting the Frequency Window

The ionosphere displays variations in electron concentration on many temporal scales; the critical frequency and the maximum usable frequency (MUF) have similar temporal variability. The eleven year solar cycle is manifested in electron density and critical frequency changes with higher values generally occurring during periods of higher solar activity. Seasonal and monthly changes are also in evidence. The consequences for HF communications performance are considerable. The operator must choose a frequency capable of sky wave support at the time of interest – in other words the operator must work below the instantaneous MUF. However, using a frequency which is too low may incur too much ionospheric absorption.

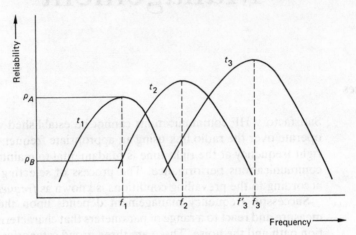

Fig. 8.1   *Time dependence of circuit reliability*

The concept of circuit reliability discussed in section 6.3.4 can be used to illustrate the effect. Consider three consecutive hours t_1, t_2 and t_3 around dawn. The corresponding circuit reliability curves for monthly median conditions might look something like those shown in Figure 8.1. The optimum working frequencies to choose, assuming that interference is not present, will be given by f_1, f_2 and f_3. (If for example f_3 happens to be corrupted by strong interference then an adjacent frequency f_3' is preferred.) The communicator must ensure that a change of operating frequency, from say f_1 to f_2, is made before the reliability of f_1 degrades substantially (it degrades from ρ_A to ρ_B between t_1 and t_2). Thus circuit reliability must not only be optimised at a specific time but should be maintained at a consistently high level over the period of circuit operation.

A number of frequency prediction programs are now available which provide monthly median predictions of optimum working frequencies for any region of the world. Link performance is evaluated in terms of signal-to-noise ratio achievement probabilities. Hence a circuit 'quality' or reliability factor (as detailed in section 6.3) can be assessed for a range of frequencies, thus defining the frequency window limits required. It is important to note that such predictions provide only monthly median estimates.

8.2.2   MUF Prediction

The starting point for the prediction of the MUF and hence of an upper frequency limit of interest is an estimate of the E and F layer critical frequencies. The worldwide distribution of f_oE, the E layer critical frequency, has a fairly simple dependence upon the solar zenith angle, local time and latitude. There is, however, no such simple dependence for f_oF2, the F2 layer critical frequency. Unfortunately it is the F2 layer with which the communicator is primarily concerned since the MUF for a given transmission circuit is related to these parameters.

Because a simple analytical representation of the temporal and local variations of the F2-MUF is not possible, the worldwide distribution of the values is given in graphical form. The primary sources of these charts are the ionograms taken by approximately 200 vertical incidence sounding stations distributed all over the world. Besides other data, hourly values of f_oF2 and MUF-factors are tabulated by these stations. Monthly medians of these values are used for constructing, after some smoothing, worldwide maps of contours of critical frequencies and MUF-factors. The resulting charts correspond only to a given month and a certain level of solar activity. Since observations covering more than one sunspot cycle have been made, there are now charts available for any level of solar activity based on either direct observation or interpolation using the established dependence of the parameters on solar activity. Most agencies engaged in prediction work currently use a method of numerical mapping of ionospheric characteristics.

The term *numerical map* is used[1] to denote a function $\phi\,(\lambda, \theta, t)$ of the three variables: latitude (λ), longitude (θ) and time (t). The function $\phi\,(\lambda, \theta, t)$ is obtained by fitting certain polynomial series of functions of the three variables to the basic ionospheric data. The numerical map is particularly useful when large numbers of propagation path computations are required.

A comparison of monthly median predictions (made using the recommended CCIR[4] prediction method for central European latitudes) with actual vertical sounding data over the last eleven year sunspot cycle shows a very reasonable agreement (to within about 10%) for the MUF; predictions are least accurate and most likely to be in error near dawn and dusk.

The temporal variability that is most difficult to model, and hence the one leading to the largest uncertainties in ionospheric propagation prediction, is the day-to-day variability. The day-to-day variability[2] of hourly values of f_oE about the monthly median hourly values, expressed in terms of the standard deviation, is around 5 to 10%. The day-to-day variability of f_oF2 is typically between 10 and 20% for both high and low sunspot activity. Observations of day-to-day variability indicate a tendency for the night-time variability, of about 20%, to exceed the day-time by about a factor of two.

The range of f_oF2 values between upper and lower quartiles of the daily values observed at a particular place and local time is[3], on average, about twice as great (and in summer as much as five times as great) as the error in the corresponding median values predicted by the CCIR[4] method. Significantly more accurate predictions of f_oF2 will therefore be obtained for most local times by including in the prediction system the day-to-day variability, rather than attempting to improve the accuracy of the predicted median values.

8.2.3 LUF Prediction

Prediction of the lower frequency limit is somewhat more complicated than the upper limit, since it depends not only upon ionospheric characteristics, but also upon equipment considerations.

The lower limit of the usable frequency range is determined by the frequency at which the signal-to-noise ratio falls below a certain value. This value depends upon type of service. (A much higher signal-to-noise ratio is, for example, required for a high fidelity broadcasting service than for hand-keyed telegraphy with experienced operators.) The calculation of signal strength in ionospheric propagation is rather complicated and not always very accurate. As a consequence, the methods proposed by various authors differ considerably.

Generally the path loss incurred by a signal that propagates over a single path consists of

a) Spatial dispersion of energy.
b) Absorption in the ionosphere.
c) Focusing or defocusing of energy by the geometry of the path.

Whilst the spatial dispersion of energy is a relatively simple calculation to perform, the absorption of energy within the ionosphere is extremely complex. Most prediction techniques necessitate the use of a 'standard ionosphere' model possessing a well behaved analytical profile of electron concentration against height. The predictions can only be as good as the model; no satisfactory methods have yet been devised to cater for horizontal spatial variations within the ionosphere which might, for example, be characteristic of crossing dawn or dusk conditions or in latitudes prone to auroral disturbances.

The calculation of propagation path loss is only part of the problem of evaluating a lower frequency limit. Whilst powerful point-to-point installations may be considered as radiating relatively constant power over the frequency band, small mobile sets with electrically short antennas may have rather different properties across the band. Therefore, any prediction of the lower frequency limit must take careful account of the transmitting antenna's performance characteristics. A poor antenna efficiency can severely limit the available frequency window, a feature explained in section 7.3.1.

8.2.4   Example Format of Long-term Predictions

Many agencies now have a variety of long-term propagation prediction programs at their disposal. These predictions supply similar information in different formats based upon a range of algorithms. Rather than describe one particular prediction program, some general comments will be made regarding the information that they contain. Consider Table 8.1 as a typical example. It contains the following information:

Line 1 contains a prediction identifier, the month and the solar activity level in 12 month moving average Zurich sunspot number.

Lines 2 and 3 contain the transmitter and receiver locations.

Line 4 defines the bearings in degrees from transmitter and receiver.

Line 5 contains the circuit range in kilometres and nautical miles.

Line 6 defines the antennas used and the transmitted signal polarisation.

Line 7 defines the transmitter output power and the 3 MHz man-made noise power level in a 1 Hz bandwidth.

Line 8 contains the hourly median signal-to-noise ratio in the specified bandwidth required for the grade of service.

1. Prediction ACB-2344 for January, SSN = 20
2. Londonderry to Washington D.C.
3. 55.00N 07.30W to 38.45N 76.51W
4. Azimuths: 280.5, 46.2
5. Range: 5333.6 km, 2880.3 NM
6. Antennas: Isotropic, Horizontal Polarisation
7. Power 200 kW; Noise −154 dBW
8. Req. S/N = 60 dB; Bandwidth 1 Hz

Time 1200	MUF 14.5	Frequencies										
		4	5	6	7	8	10	12	15	17	20	
		3E	4F	3F	2F	2F	2F	2F	2F	2F	-	Mode
		3	15	10	4	3	3	3	4	4	-	Angle
		181	190	187	184	183	183	183	184	184	-	Delay
		99	99	99	99	99	99	90	39	8	-	C. Prob
		5	53	67	75	81	91	93	101	103	-	S/N dB
		0	14	67	84	93	98	90	39	8	-	Rel

Table 8.1   Example of a long-term prediction

There then follows a prediction for 1200 GMT showing the circuit MUF to be 14.5 MHz. For each operating frequency the body of the tabulation contains

MODE   The propagation mode having the greatest probability of occurrence.

ANGLE   The median elevation angle in degrees associated with the mode.

DELAY   The propagation time in tenths of milliseconds.

C. PROB   The percentage of days that the sky wave mode is expected to exist (c.f. section 6.3.3).

S/N dB   The median of the hourly median signal-to-noise ratio in dB for the days the sky wave mode exists.

REL   The percentage of days within the month that the required signal-to-noise ratio is expected to be equalled or exceeded (c.f. section 6.3.4).

Other definitions and information are provided in different prediction methods but the basic output formats are very similar.

8.2.5   Limitations of Long-term Predictions

Long-term predictions do not aim at predicting propagation conditions at a given date. Rather they correlate sky wave propagation to given levels of solar activity. The establishment of the correlation is based on worldwide observations in the past, covering at least one full sunspot cycle. The limiting factor is the completeness of the observing network and the accuracy of the conversion of vertical to oblique incidence data (see section 8.3.2).

Long-term predictions provide broad guidance as to the expected frequency band or window which should be used, as a function of communications range, time of day, season and solar cycle. Such predictions will be valid for 'normal' conditions only.

It has not yet proved possible to predict the likely occurrence of sporadic E. On occasions, the critical frequency in the presence of sporadic E is much greater than for the normal conditions when reflections are returned from the F2 layer. The likely effect on onset of sporadic E conditions is a rapid and considerable increase in the available frequency window.

The problems inherent in high latitude propagation paths are such that there are still no satisfactory models or forecasts that have any practical usefulness. For example, the onset of ionospheric storms implies a depression of the MUF and an increase in D-region absorption. These two features together imply a narrowing of the frequency window. High latitude ionospheric propagation represents the greatest challenge to the system designers and operators of improved HF links.

The following points can be made regarding long-term predictions:

a) CCIR numerical map methods of MUF prediction appear to be superior to other, simpler, algorithms.
b) Prediction methods which use CCIR maps are more likely to be in error during the winter than the summer.
c) Prediction errors vary with local time and the greatest errors often occur near dawn and dusk.
d) Prediction errors increase with latitude. They may be considerable at high latitudes.
e) The accuracy of long-term predictions could be improved by incorporation of better ionospheric maps into the prediction method. Current maps are based upon very sparse information in certain areas (for example in the Arctic, Antarctic, Pacific and Sahara regions). To produce better maps from raw sounding data in such regions would be a major undertaking.

8.2.6   Short-term Predictions

Short-term predictions are used to update the long-term predictions in the light of currently prevailing conditions and to cater for any anomalous propagation mechanisms that might be present. Short-term updates are useful in that the circuit operator may be able to anticipate MUF failure and thus maintain the circuit reliability.

Information availability on short-term ionospheric variability can be of critical importance. Short-term predictions are provided by bulletins issued by short-term forecasting/disturbance warning centres. Daily forecasts provide indications as to the expected circuit quality relative to monthly median conditions, for example 'Below normal radio conditions are expected for the next 24 hours' or 'Northern circuits may become increasingly disturbed'. The experienced radio operator can interpret this information in terms of the required changes to the 'undisturbed' operating frequency.

An example of a propagation forecast service is provided by the Radio Station WWV at Fort Collins, Colorado which broadcasts on 2.5, 5, 10, 15, 20 and 25 MHz. Announcements of short-term forecasts for propagation along paths in the North Atlantic area are made at 14 minutes past each hour, whilst geophysical alerts and solar flux forecasts are broadcast at 18 minutes past each hour.

Another, more localised, method of obtaining short-term forecasts is by the use of a vertical sounder (see section 8.3) co-sited with the radio equipment. Data from this can provide evidence of any anomalous propagation conditions. It can be used to update the long-term predictions accordingly. Such information would, of course, need to be interpreted in terms of oblique paths of interest to the communicator. However, the usefulness of the current systems is limited, amongst other things, by the area of coverage of ionospheric sounding data, the speed with which measurements are reported and the time taken to distribute the final forecast[5].

In order that short-term forecasts can be compiled and issued within, say, one hour, each sounding station needs to be able to report details of its ionogram to the forecasting centre. Data-gathering on this scale could be accomplished using land lines or meteor burst communications techniques. Future sounding equipments are likely to incorporate storage facilities whereby a user may interrogate sounding information on demand. Alternatively, the information could be broadcast by HF.

8.3 Sounding

8.3.1 Types of Technique

Ionospheric sounding is here used to refer to sounding the propagation medium characteristics such as channel unit impulse response, signal propagation delay and signal amplitude. By contrast channel evaluation, discussed in section 8.4, provides information for use in association with signal-to-noise such as data error rate, speech intelligibility and noise levels.

Although the primary aim of the various sounding methods is to provide propagation information, some techniques also provide the opportunity to evaluate signal-to-noise performance. Hence the distinction between sounding and channel evaluation is not always delineated clearly.

The following techniques, each of which is discussed in more detail in subsequent sub-sections, have been developed for sounding:

1 *Ionospheric Pulse Sounding*
This is concerned with the measurement of the linear unit impulse response function for each channel. Measurements may be made over either a vertical or an oblique path. The output of the sounder is usually displayed in the form of an ionogram. One extension of this basic technique is modulated pulse sounding; individual pulses on each channel are modulated by a digital waveform, which can be used to transmit information, for example, noise levels in assigned channels. In this way it may be considered very similar to channel sounding, described in **3**.
2 *Linear Sweep Sounding or 'Chirpsounding'*
This technique uses a linear frequency swept signal as the sounding transmission. The output can be displayed as an ionogram[1].

3 *Channel Sounding*

This technique makes soundings only on a discrete number of allocated channels. The sounding signal may additionally include information relating to interference levels. This technique can be extended to include interference levels of the sub-channels in the case of FSK transmissions. It may also be desirable to assess a channel while it is actually carrying traffic in order to determine when a frequency change becomes necessary. This may necessitate modification to the transmitted signal to allow the appropriate information to be extracted. The incorporation of interference and channel assessment information provides the basis of real-time channel evaluation (see section 8.4).

8.3.2 Pulse Sounding and Ionograms

During pulse sounding powerful pulses of short duration are transmitted at preselected frequencies and bandwidths. The received signal is then analysed. Three basic sounding schemes are used (see Figure 8.2):

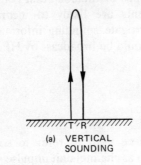

(a) VERTICAL
     SOUNDING

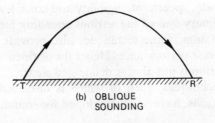

(b) OBLIQUE
     SOUNDING

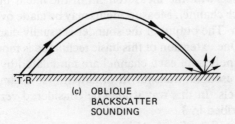

(c) OBLIQUE
     BACKSCATTER
     SOUNDING

Fig. 8.2 *Basic sounding techniques*

1 *Vertical Incidence Sounding*
The sounding pulse is emitted vertically and the reflected returns from the ionosphere are analysed at a co-located or nearby receiver (Figure 8.2a).
2 *Oblique Incidence Sounding*
The sounding pulse is emitted either over the actual communications path, or over an adjacent path (Figure 8.2b). This method requires the transmitter and receiver to be remotely synchronised.
3 *Oblique Incidence Backscatter Sounding*
As with the vertical sounding the transmitter and receiver are co-located or relatively close, but the received signals are reflected obliquely from the ionosphere and are scattered from ground irregularities (Figure 8.2c).

When the transmitted pulse is of short duration, the sounder receiver measures the linear unit impulse response function for each channel. The output from the sounder is usually presented as an *ionogram*, which is a two-dimensional projection of the raster of impulse responses for each channel.

Vertical and oblique incidence ionograms have substantially different appearances. Consider as a simple example, ionograms produced for a single

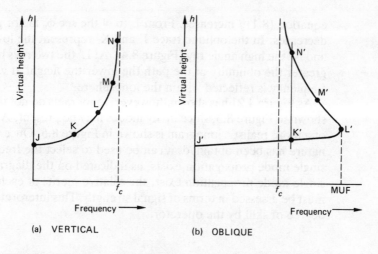

(a) VERTICAL (b) OBLIQUE

Fig. 8.3 *Comparison of vertical and oblique ionograms*

propagation mode as shown in Figure 8.3. The letters J, K, L, M, N denote points on the vertical ionogram. For a plane ionosphere

$$f = f_v \sec \phi_0 \qquad (8.1)$$

[see equation (4.5)] where ϕ_0 is the angle of incidence on the ionosphere by the oblique ray of frequency f; f_v is the equivalent vertically incident frequency. The equivalent oblique ionogram is also plotted in Figure 8.3. The vertical scale in Figure 8.3b can be related to the equivalent oblique path by the transformation $P' = 2h' \sec \phi_0$ [equation (4.7)]. As the frequency is increased from J in Figure 8.3a the virtual height increases and $\sec \phi_0$ decreases. $\sec \phi_0$ decreases more slowly than f_v increases between J and L, and hence from

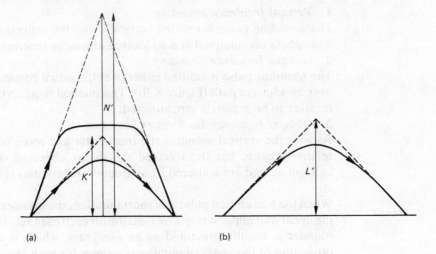

Fig. 8.4   *Ray paths associated with an oblique ionogram*
(a) High and low angle ray paths
(b) MUF ray path

equation (8.1) f increases. From L to N the sec ϕ_0 factor predominates and f decreases. In the oblique trace J' and K' represent the low angle ray and M' and N' the high angle ray (Figure 8.4). At L' the two rays have coalesced. The greater the obliquity of the path the lower the height at which the maximum frequency is reflected within the ionosphere.

As Figure 4.21 has shown, however, ionograms do not take the simple form shown in Figure 8.3, since many modes may be present. A schematic example of a more realistic ionogram is shown in Figure 8.5. Once an ionogram of this nature has been obtained, it can be used to select the frequency range where single mode propagation exists, as indicated on the diagram. If no regions of single mode propagation exist, the relative merits of each propagating mode must be assessed in terms of signal strength. This interpretation requires some degree of skill by the operator.

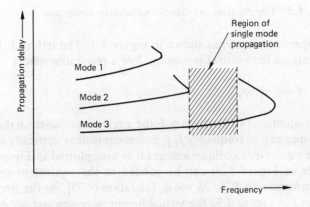

Fig. 8.5   *Interpretation of oblique ionogram*

8.3.3   Comparison of Sounding Techniques

If the sounding pulses are modulated by a digital waveform, the following advantages can be gained[7]:

a) Pulse compression coding can be applied[9]. This improves the delay resolution properties of the system without having to shorten the pulse length and hence increase the peak transmitted power to maintain the same transmitted energy.

b) Small quantities of data can be transmitted with the sounding signal, via pulse modulation coding. This can be used, for example, to convey noise levels in assigned channels.

A single vertical ionosonde station will provide information on ionospheric variability, not only at the station location, but also within its surrounding area. The shape and extent of this area depends on the spatial correlation of f_oF2 variability and the degree of forecasting improvement required. To halve the f_oF2 prediction uncertainty may require a vertical incidence sounding network with station spacing no more than 500 km in the north-south direction and 1000 km in the east-west direction. On this basis, between 12 and 15 sounders would cover the area of northern Europe, for example. To reap the full benefit of a vertical sounding network, the measurements must be collected, processed and a forecast issued to operators within an hour or so.

Both oblique incidence and oblique incidence backscatter sounding techniques require broadband directional antennas. The propagation path is better defined by the former, but information is only available on a limited number of paths. The latter gives greater flexibility of paths which can be sounded, but at the expense of requiring a more powerful transmitter due to the greater distance the sounding signal must travel.

A typical pulse sounder operates with a peak transmitter power of 30 kW, using a pulse of about 1 ms duration with 2, 4 or 8 pulses per channel at a rate of 20 pulses per second. More than one hundred channels can typically be sounded.

The strength of pulse sounding lies in the ability to utilise a single network to provide information to many users worldwide. There remains the problem of collecting together the available information and transmitting it to the appropriate location in sufficient time for it to be useful.

The relative merits of the sounding techniques may be summarised as follows:

1   With vertical and oblique backscatter sounding the transmitter and receiver are co-located (or relatively close) simplifying the synchronisation of the two units. However, with oblique sounding, the transmitter and receiver are remote.

2   With oblique sounding a single receiver can be used to construct ionograms from a number of remote transmitters. This eases the problems of bringing together propagation conditions quickly. The same naturally applies to transmitters, giving rise to a flexible sounding network.

3   Oblique backscatter sounding can measure propagation conditions for a number of paths, but at the expense of greater transmitter power requirements due to the longer propagation path.

4   Oblique backscatter sounding gives rise to a number of possible returns for a given path, corresponding to all elevation angles of propagation resulting from scatter from different ranges. Thus, for each frequency, a number of backscatter returns will be received, beginning with the minimum time delay return. The leading edge of the multiple returns is often the only useful portion of the backscatter ionogram, since all other returns cannot be distinguished from one another without very high resolution ionograms.

8.3.4   Linear Sweep Sounding (Chirpsounding)

Linear FM modulation or chirpsounding[10] consists of sending a low power 2–30 MHz linear FM/CW test signal over the communications path. The same antenna as for the communications system can be used to take into account antenna patterns. A system block diagram is shown in Figure 8.6. The system consists of three units of equipment in addition to the communications transceivers: a chirpsounder transmitter, an associated receiver to measure propagation conditions and a spectrum monitor to measure interference and noise conditions.

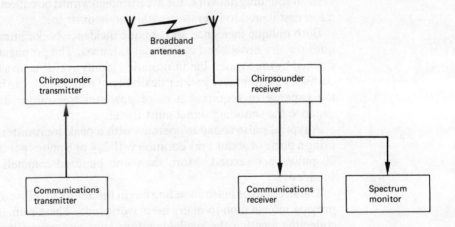

Fig. 8.6   *Simplified diagram of chirpsounding system*

Figure 8.7 illustrates the principle of the technique; in a multipath propagation situation several weighted versions of the transmitted sweep will be received as shown. If a correctly timed local oscillator sweep is available at the receiving site, this can be mixed with the incoming sweep components to yield the difference frequency components, which are then subjected to spectral analysis.

After mixing with the local oscillator signal, the frequency components f_1, f_2, f_3 of the difference signal are $\Delta t_1 \, df/dt$, $\Delta t_2 \, df/dt$, $\Delta t_3 \, df/dt$. Hence propagation delays are translated directly into frequency offsets.

Chirpsounding is used not only to sound the communications path obliquely,

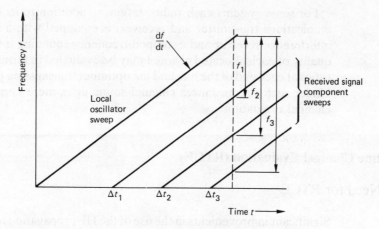

Fig. 8.7 *Principle of chirpsounding*

but also for vertical sounding. This provides ionograms of quality equal to or better than those provided by pulse sounders, with the advantage of causing less RF interference to nearby equipment.

The sounding signal is tracked by the time synchronised receiver. Spectral analysis of the difference frequency between the receiver local oscillator and the incoming signal yields an *oblique chirpsounder record*, i.e. a time delay versus radio frequency display. Received power versus frequency may also be displayed. The spectrum monitor is used to compile the channel occupancy statistics for a number of channels in the HF band. Statistics are compiled on the interference levels of each channel and stored. A channel is defined as unoccupied if no interference is present during an operator definable period of time. A decision on the optimum frequency for communications can be determined from the results of the two displays, i.e. the frequency range giving the best propagation conditions and the channels within that range providing the lowest interference levels.

It may be desirable also for the receiver to have knowledge of the interference levels or status information at the transmitting station. This can be accomplished by modulating simple messages on to the chirpsound signal. However, the scope for such information transmissions is somewhat limited.

8.3.5 Channel Sounding

Channel sounding techniques employ probing transmissions on only a limited number of allocated frequencies over the radio link of interest. Both ends of the link must be maintained in time and frequency synchronisation; this is generally achieved by crystal-controlled clocks. Since only allocated frequencies are 'sounded', time is not wasted by sounding channels that are unavailable for communications, as occurs with swept sounding. Advantage can be taken more readily of unusual propagation conditions such as sporadic E although, of course, a full picture of ionospheric conditions is not obtained.

For some systems each radio station, in addition to its conventional communications transmitter and receiver, is equipped with a stepped-frequency interference receiver and a stepped-frequency sounding transmitter. The link quality of each allocated channel may be evaluated in terms of signal-to-noise ratios at each end of the link and the optimum transmission frequency selected. Under such circumstances channel sounding is more correctly described as channel evaluation.

8.4   Real Time Channel Evaluation (RTCE)

8.4.1   The Need for RTCE

Significant improvements in the use of the HF propagation medium can only be achieved if an operator, or an automated HF system controller, has access to real-time data on the relevant path parameters rather than having to rely on off-line propagation analysis. Since it is the occurrence of interference rather than the propagation conditions that will often limit system performance, *channel evaluation schemes* can be particularly effective in improving HF channel performance. This interference is found to be at its highest during the night and at dusk, less at dawn and at frequencies near the optimum traffic frequency (FOT) during the day, and least at frequencies well below the FOT during the day. Before discussing the nature of real time channel evaluation (RTCE) it is useful to review the characteristics and shortcomings of the prediction methods discussed so far.

The aim of off-line propagation analysis procedures is to provide frequency selection data which will give the communicator a 90% probability of satisfactory communications at any time, assuming that the basic characteristics of the communications system, for example transmitter power and antenna design, have been correctly specified in the system design. Typically, the frequency selection data would be provided in the form of an optimum working frequency on an hour-by-hour basis. The classical method of controlling an HF circuit using off-line propagation analysis data is to select one or more daytime operating frequencies and one or more night-time frequencies, as shown in Figure 8.8. Although propagation analysis programs are being refined continuously, they have certain fundamental limitations which lead to off-line predictions being less than adequate for many purposes. The most important of these shortcomings are

a) The effects of interference from other spectrum users are not included in the analysis and prediction model.
b) The propagation data base used for computation of predicted circuit performance is limited.
c) The effects of significant perturbations such as sudden ionospheric disturbances (SIDs), ionospheric storms and polar cap events (PCEs) cannot, by their random nature, be taken into account in the analysis.
d) The effects of relatively transient propagation phenomena, such as sporadic E layer refraction, can be described only approximately.
e) The 'confidence level' for the predictions is normally only 90%.

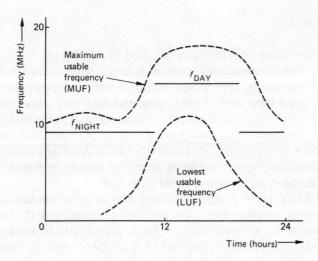

Fig. 8.8   *Example of two-frequency operation*

To overcome some of the above limitations in long-term ionospheric forecasting techniques, short-term forecasting techniques have also been developed. These involve real-time observations of solar and ionospheric parameters, together with feedback of information concerning which frequencies are propagating at a given time on selected circuits. Clearly, these procedures will tend to overcome some of the limitations of the long-term forecasts; however, the following points should be noted:

a) Correction data can be provided only on the basis of sampled real-time conditions and therefore will not be uniformly accurate for all links.
b) There are logistic and economic problems associated with the timely dissemination of the correction data.
c) In general, the corrections do not indicate which of a set of assigned channels is likely to provide the best grade of service.
d) As with long-term forecasting techniques, the effects of man-made interfering signals are not indicated.

For the reasons listed above, off-line propagation analysis cannot normally provide circuit parameter forecasts with a degree of confidence required for HF communications where high reliability and availability are essential. Accordingly, increasing emphasis is being given to methods for characterising HF channels accurately in real time. The need for RTCE is most pronounced for links involving mobile terminals since the characteristics of the paths will change with time thus making off-line analysis more approximate.

8.4.2   The Nature of RTCE

A definition of RTCE now adopted by CCIR[6] is

'Real-time channel evaluation is the term used to describe the processes of measuring appropriate parameters of a set of communications channels in real time and of employing the data thus obtained to describe quantitatively the states of those channels and hence their relative capabilities for passing a given class, or classes, of communications traffic.'

The RTCE process[7] is essentially one of deriving a numerical model for each individual channel in a form which can readily be employed for performance prediction and system control purposes.

RTCE is not concerned simply with an up-to-the-minute assessment of HF propagation conditions, but also with characterising the effects of interference from other spectrum users. This is particularly important because, in many instances, it is interference which is the factor limiting communications system performance, rather than propagation.

The term 'real time' implies that the measured channel parameter values are updated at intervals which are of the same order as the overall response time of the communications system to control inputs. There is no advantage in making measurements more frequently since the information cannot be employed effectively by the communications system. A channel model must be generated in a form appropriate to the type of traffic which is to be transmitted. For example, a channel model required for 75 bit/s telegraphy would normally be expected to be considerably less complex than that for a 2.4 kbit/s digitised speech link.

An efficient RTCE system will provide much more information than simply the optimum frequency for transmission. Recommended start times and data rates should ideally be provided. Use can also be made of modes other than the more conventional ones; for example the RTCE system should take advantage of sporadic E effects when they are advantageous to the communications channel performance.

8.4.3   Automation of System Control

If manual control procedures are used, response times will be, at best, a few tens of seconds and there will be no chance of the system being able to utilise relatively short duration, high capacity transmission windows (analogous to the meteor burst situation). RTCE, however, is able to provide the necessary information to enable the system control procedures to be automated – hence potentially providing a greatly improved response time. Indeed, RTCE is an essential prerequisite for the application of automatic system control procedures. The output from the RTCE process must therefore be expressed in terms which are meaningful to the automated system controller, for example in terms of a predicted error rate for digital data transmissions.

In the HF band, high radiated power levels are undesirable. Not only do they require an increase in transmitter and antenna sizes, and hence costs, but they also exacerbate spectral pollution. The use of RTCE in an automated HF communications system will tend to reduce the necessity for higher radiated powers by selecting frequencies and transmission times for which the received signal-to-noise ratio is maximised and interference avoided. The importance of noise and man-made interference in determining HF communications system performance has already been stressed. In areas of high spectral congestion such as central Europe, it is normally man-made interference which limits system performance, rather than propagation, which is often relatively predictable. It is clear, therefore, that considerable effort must be placed upon the measurement and characterisation of interference. As a general principle, system availability and reliability should be improved by the use of RTCE and more effective signal processing, rather than by transmission at higher power levels. Automatic HF system control requires the application of considerable information processing power. Therefore, future HF communications systems will inevitably be processor-based, with manual intervention only in exceptional circumstances.

The RF stages of the communications system should be capable of rapid frequency agility so that full advantage can be taken of automatic and adaptive system control procedures. The ability to change frequency rapidly will also facilitate the RTCE process to find a channel with an acceptable signal-to-noise ratio. This will eliminate the 'resistance' of system operators to changing frequency except where absolutely essential and thus lead to improved flexibility and lower radiated power levels.

8.4.4   Matching the Channel to the Medium

RTCE techniques are designed only to monitor and select the best of a set of alternative transmission channels at a specified time; in general, they are not well matched to the task of monitoring the short-term time variability of those channels. Thus, current RTCE algorithms search for transmission windows which are likely to persist for the complete duration of a fixed constant rate transmission; at the same time, they attempt to minimise the long-term overlap between communications and noise/interference windows.

There is a degree of similarity between HF channels and meteor burst paths. Communications using ionised meteor trails can take place only for short periods of time (typically $0.5 - 1$ s) when a usable trail occurs; the time between the occurrence of such trails is relatively long, tens of seconds or more. However, when the path does exist high rate transmission is possible since the propagation is essentially single mode. The HF medium also gives rise to high capacity windows, but in contrast to the meteor burst path, having zero capacity between windows, the HF path has a residual, lower and variable capacity between windows.

The majority of existing HF systems[8] operate at a constant transmission rate with fixed bandwidths and signal formats. This situation forms a fundamental mis-match between the techniques employed and the essential nature of the propagation medium. A constant rate HF system operates at well below its potential capacity for much of the time. An improvement could be made by use of an adaptive system in which parameters can change in response to changes in available capacity. However, even if constant rate transmissions continue to be used, the source and channel encoding procedures could still be made adaptive to counter changes in the received signal-to-noise ratio, as described in section 10.5.

8.4.5   An Air-Ground Link Example

To illustrate some of the concepts of real time channel evaluation consider the example of an air-ground link.

Channel evaluation schemes operate on the basis of a number of frequency blocks (see Figure 8.9) spread across the HF band. Each block is less than a few hundred kilohertz wide, which is somewhat less than the bandwidth over which propagation conditions usually could be expected to be reasonably constant. Thus, an evaluation of propagation made at one frequency within a block can be assumed valid for any other frequency within that block. Probing signals or soundings need therefore only to be made on one of these frequencies which will be dedicated for channel evaluation. The remaining frequencies are then available for communications.

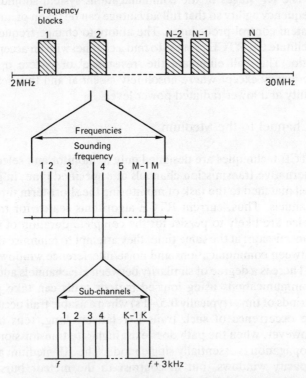

Fig. 8.9 *Frequency distribution in a channel evaluation scheme*

For an air-ground system, soundings sent from a ground station contain information to be used by the aircraft when replying to, or initiating, calls. There are two basic types of transmission:

a) The evaluation or sounding message (ground-to-air only).
b) The handshake message (both ground-to-air and air-to-ground).

The amount of error control coding chosen for each of these transmissions must be sufficient to give a good performance over the type of communications links of interest.

In addition to receiving and analysing sounding signals the system must measure the level of interference on all allocated frequencies within the blocks, i.e. the sounding frequency and all associated communications frequencies. These measurements are used to rank the frequencies within a block in terms of interference levels for different types of traffic. Thus, for each block, the ground-to-air traffic frequency within a frequency block can be chosen using the frequency with the lowest overall interference for voice communications or the one with low sub-channel interference for low rate telegraph transmissions.

Each available channel is evaluated in turn. The first part of the process for each channel consists of a measurement of background interference level at the ground station. This level is then encoded and transmitted to the aircraft. If propagation is possible between the ground station and aircraft, these encoded transmissions will be detected by the airborne receiver and provide the aircraft with the interference level at the ground station. This transmission will also be used to measure the signal-to-noise ratio of the channel at the aircraft end of the link. In this way the aircraft can build up a picture of the interference levels at each end of the link for each channel, and hence select the optimum channel for communications with the ground station when desired.

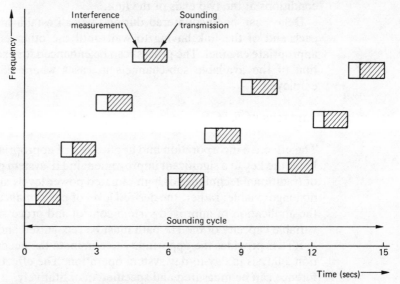

Fig. 8.10 *Example sounding cycle*

Figure 8.10 shows a simple example for 10-channel operation. If the interference evaluation process on each channel takes 0.5 second and the sounding signal is of one second duration then the total *sounding cycle* will be of 15 seconds duration. During this cycle time the aircraft will have built up a complete picture of

a) Propagation conditions – from signal strength measurements of the soundings.

b) Ground station noise levels – from information supplied in the soundings.

c) Aircraft noise levels – from measurements made locally when no soundings are being transmitted.

When the aircraft wishes to call the ground station, it selects what it believes to be the optimum channel and transmits a coded message on this channel at the time the ground station is monitoring that channel for interference levels. This coded message contains the channel to be used for ground-air communications. The ground station responds on the appropriate channel and receipt of this is acknowledged by the aircraft. Handshake is thus established and communications can then take place. Should this procedure fail at any point, alternative frequencies will be selected and the process repeated.

When the ground station wishes to call an aircraft, it includes the aircraft 'address' in the probing transmission. This may have to be repeated on several channels as the ground station has no knowledge of the propagation conditions to the aircraft. The aircraft subsequently replies on what it considers to be the optimum channel.

Such a system often selects channels outside the frequency range which would be expected from forecasts, suggesting that more spectrum space is available than can be utilised using non-real-time frequency management techniques. On a significant proportion of occasions, different frequencies might be selected for the two directions because of different interference conditions at the two ends of the link.

Delays can be encountered during this call establishment process, due to each end of the link having to wait until the other end is monitoring the appropriate channel. The process can be enhanced for example, by the evaluation of the available sub-channels in cases where FSK transmissions are employed.

8.4.6   Potential Advantages of RTCE

The effective incorporation and application of appropriate RTCE techniques holds the key to a significant improvement in HF system performance. The use of traditional techniques of high radiated power levels and large antennas are no longer viable; rather, the desired level of performance should be sought by the application of improved system control and processing. In particular the variable capacity of the HF path must be recognised and exploited.

RTCE provides the opportunity to eliminate the need for off-line propagation analysis in day-to-day system operation. The effects of man-made interference can be measured and specified quantitatively.

The availability both of frequency channels higher than predicted and of relatively transient propagation modes can be identified and exploited by RTCE. Not only will this provide a better communications service but will assist in reducing spectrum congestion by using frequencies higher than normally expected to be of value.

Spectral pollution can also be minimised by adaptive control of radiated power. RTCE provides the basic data required for adaptation of system parameters other than frequency, for example antenna characteristics and signal processing algorithms.

9 Data Communications

9.1 General Considerations

The discussion thus far has not explicitly concerned itself with specific issues of data communications. However, this is an extremely important aspect of HF communications and will become more so in the future. Traditionally, HF has provided a voice and low-speed telegraphy capability. Current requirements are for data transmission at rates of typically 2400 bit/s to support computer data links and secure digital voice. Digital communication, however, still operates over 3 kHz voice channels. This chapter is devoted to aspects of data transmission that impact upon the HF communications channel.

Many circuits employing data transmissions have a large frequency complement. This allocation can be used to advantage to choose frequencies close to the maximum usable frequency (MUF) and thus ensure that differential delays between propagation modes are small enough to provide frequency flat fading (i.e. independent of frequency) over a 3 kHz channel. Post-detection diversity combining can be employed to combat such fading by using spaced receiving antennas and multiple transmission frequencies. The digital errors that remain are then caused predominantly by either wideband impulsive noise or man-made interference; the time-varying dispersive effects of the channel are of secondary importance. In principle, therefore, the performance of these point-to-point links may be optimised by good engineering design and practice in respect of the equipment and antenna systems, whilst high transmitter power is often available.

Much more difficult problems are presented by HF communications to mobiles. When transmitting data to and from these mobiles, it is neither easy nor always possible to use frequencies close to the MUF as in the case of point-to-point links with large frequency complements; time-varying channel dispersive effects can then become of primary importance. At the frequencies available to the mobiles, the resulting differential delays between propagation modes may be sufficient to produce narrowband frequency selective fading within a 3 kHz channel. To achieve satisfactory results over a HF link of this kind careful consideration must be given to the HF channel characteristics, the terminal radio equipment (including modulation techniques and error control coding), the planning of operational links and the management of the frequencies to be used over those links.

Further complications arise when high data rate transmissions are required. For example, digital voice requirements imply data rate transmission of 2.4 kbit/s. Higher data rates would give better quality from the speech synthesis aspect, but channel bandwidth considerations show that approximately 2.4

kbit/s is the highest rate that can be tolerated in a 3 kHz channel. Military radio links may need to incorporate a high degree of immunity to electronic counter measures; complex modulation schemes involving frequency hopping and spread spectrum must therefore be adopted. This, in turn, necessitates a detailed study of the propagation medium to determine whether various forms of wide bandwidth modulation techniques can be transmitted with fidelity. The problems inherent in the design of modems to achieve satisfactory transmissions at 2.4 kbit/s over HF channels have not yet been adequately solved; it has been the ionosphere which has proved to be a limiting factor in the design of an efficient modem.

The main task in designing digital communications systems is to find the appropriate modulation technique which can operate at the highest possible speed. In any communications channel with a bandwidth of B Hz, the maximum data rate that can be transmitted and received with no error was found by Shannon[1] to be given by

$$C = B\log_2 (1+S/N) \tag{9.1}$$

where C is measured in bits per second and S/N is the band-limited signal-to-noise ratio.

However, transmission speed alone is not enough, since errors may arise in the reconstruction of the data at the receiver. What is required is a means for transmitting a maximum volume of data within a given limited bandwidth. This must be received reliably and decoded with minimum error at a reasonable price.

9.2　Digital Modulation Techniques

9.2.1　Terminology

In a digital data communications system the information source consists of a finite number of discrete *symbols* which are coded into a sequence of *waveforms*; each waveform is selected from a finite alphabet of signal waveforms. Thus the requirement to transmit information is reduced to the problem of transmitting a sequence of waveforms, each one selected from a specified and finite set. This is in contrast to the problem of transmitting analog information where the resulting set of waveforms is infinite.

The conversion process of digital data to an analog signal is performed by means of modulation in which a wave of constant amplitude and frequency is changed according to the data value, by shifting its amplitude, frequency or phase. The modulated signal, if all three parameters A_m, ω_m and θ_m shift the carrier, may take the form:

$$S(t) = (A_c + A_m) \cos [(\omega_c + \omega_m + \theta_m)t] \tag{9.2}$$

where A_c is the constant amplitude and ω_c $(= 2\pi f_c)$ refers to the carrier's fixed frequency. Generally only one of the parameters is varied in each modulation technique. The resulting data stream has the quality of being normally at some constant value for a period of time and then being subject to an instantaneous transition to another level and remaining there for a period. For this reason data modulation is better known as *data shift keying*. Keying is a form of modulation process which involves the selection from a finite set of discrete states.

In the detection process, either *coherent detection* (where the receiver is phase-locked to the transmitted carrier wave) or *non-coherent detection* (where the receiver is not phase-locked with the transmitter) is used. At the receiver the problem of reception reduces to the problem of deciding between the signal waveforms. Since the decision of the discrete signal receiver is either right or wrong, the criterion of performance of a digital communications system is ordinarily based on the probability of error, i.e. the probability of choosing an incorrect symbol from a finite set of possible transmitted symbols.

It is possible to divide the variety of modulation schemes into a number of different types, for example, constant envelope, non-constant envelope, continuous phase, discontinuous phase, phase modulation, frequency modulation or amplitude modulation. The boundaries are vague, however, and each type can be considered as a special case of a more complicated scheme. In the past it was normal to separate the modulation technique from the coding technique and to consider the merits of each independently. This, too, is not strictly appropriate since the newer modulation schemes combine the coding with the modulation.

For the purposes of this discussion modulation schemes are considered primarily from the HF viewpoint and are divided into two types:

a) Those that maintain phase continuity over the symbol boundary.
b) Those that can produce phase discontinuity at the symbol boundary.

For a more detailed treatise of modulation the reader is referred to one of the many general communications systems texts[24].

9.2.2 Phase Discontinuous Schemes

These are generally the older modulation schemes and can be divided into those modifying the amplitude, the phase or the frequency of the carrier.

1 Amplitude Shift Keying ASK

Shift keying the carrier signal between amplitude levels in accordance with the finite set of states is known as amplitude shift keying (ASK).

Often in binary modulation systems the carrier is pulsed so that one of the binary states is represented by the presence of the carrier while the other state is represented by its absence (see Figure 9.1). On-off keying (OOK) as it is known, involves a penalty of 3 dB. This arises because of the need to avoid exceeding the peak power limits of the transmitter.

For incoherent demodulation, known as *envelope detection*, signal phase coherence is not required in the detection process but there is often a problem in establishing a suitable reference level for the threshold detector. Contamination by interference, which would be unlikely to be equally spread over the

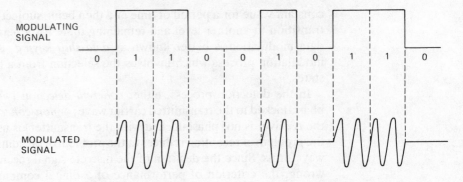

MODULATING
SIGNAL

0 1 1 0 0 1 0 1 1 0

MODULATED
SIGNAL

Fig. 9.1 *Amplitude modulation of a binary pulse stream (on-off keying)*

signal bandwidth, makes assessment of the mean level by integration over the signal bandwidth very risky. The best technique may be to use[2] a *law assessor* to establish an average signal level for each tone. Developed to combat frequency selective fading of multitone systems it establishes the presence or absence of a signal by comparing the signal energy in each bit with the mean power over a number of bits. For typical signalling rates of 50 to 75 Hz, the integration period is approximately 1/5 sec, i.e. 10 to 15 bits.

2 Phase Shift Keying PSK

In digital phase modulation, digital information is transmitted by using as a code the sequential transmission of carrier pulses of constant amplitude, angular frequency and duration but of different relative phase.

Phase shift keying involves switching the phase of a carrier between a number of discrete phases. Binary phase shift keying, BPSK, has two levels corresponding to the two possible data levels. These are normally 180° apart (see Figure 9.2). Coherently detected BPSK, CBPSK, is a simple and efficient modulation scheme. Because of phase ambiguities during demodulation, bit polarity is not maintained through the system. The fixed reference detection is susceptible to phase jitter and other channel disturbances. This creates the

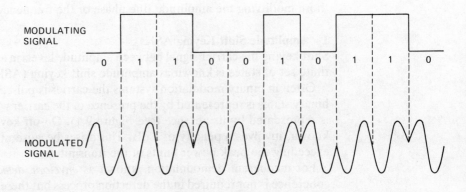

MODULATING
SIGNAL

0 1 1 0 0 1 0 1 1 0

MODULATED
SIGNAL

Fig. 9.2 *Phase shift keying of a binary pulse stream*

problem of both establishing the fixed reference and maintaining it in its fixed phase. Once the receiver loses synchronisation it takes time before the synchronisation can be re-established. Where this is important the data is normally differentially encoded and decoded with respect to a fixed reference standard to give Differentially Encoded Binary Phase Shift Keying, DEBPSK. The preservation of bit polarity is at the expense of degraded performance because the differential decoder doubles the received error rate.

3 Differential Phase Shift Keying DPSK

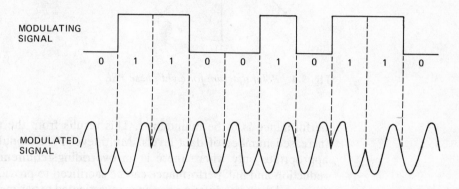

Fig. 9.3 *Differential phase shift keying of a binary pulse stream*

This differs from differentially detected CBPSK in that the demodulator does not extract a carrier reference, but uses a one bit received version of the signal as a demodulation reference. Its theoretical performance is worse than for coherent detection, but is simpler to implement. Figure 9.3 shows its operation. The information is conveyed by the change in phase between one element and the next (e.g. binary 0 = no change; binary 1 = phase reversed). Since the phase of the reference sinusoid is determined by the received phase of the previous element, this reference phase may be disturbed by noise, particularly on a fading signal. The result is that errors tend to occur in pairs so the performance is poorer than coherent PSK by about 3 dB on fading signals. However, DPSK does have advantages for HF, as discussed in section 9.2.4.

4 Multi-phase Shift Keying MPSK

Increasing data transmission speed can be achieved by increasing the number of modulation states. Multi-amplitude levels are susceptible to Gaussian noise so that, for systems with noisy environments, phase shifts could be more advantageous. Digital modulation schemes employing multi-phase shifts are referred to as M-ary PSK. The signal amplitude remains constant for all transmitted modulation states. Figure 9.4 shows an example of eight phase M-ary PSK where a tribit (3 bits) is transmitted for each phase change. Thus in MPSK each symbol represents more than one bit of information, resulting in a symbol rate that is lower than the input data rate. For MPSK, data is then transmitted in a reduced bandwidth at this lower symbol rate. The penalty for bandwidth saving is an increased energy requirement per bit if error rate

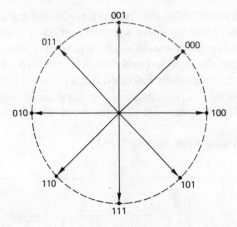

Fig. 9.4   *Polar diagram for eight-phase PSK*

performance is to be maintained. This results from the need to resolve an increased number of data levels. Multi-level PSK modulation schemes are appropriate only where there is an overriding requirement for bandwidth reduction and link performance can be sacrificed to provide this. In the USA 6-ary and 8-ary modulation has been investigated experimentally for 4.8 to 9.6 kbit/s HF adaptive modems.

Quadrature phase shift keying, QPSK, is a 4-level PSK and is equivalent to the superposition of two orthogonal BPSK channels. Four phases are possible, each 90° apart. QPSK has a bit error rate performance identical to BPSK, but occupies half the bandwidth for a given data rate. Differential quadrature phase shift keying, DQPSK, is a method of detecting a differentially encoded QPSK signal; it has a poorer noise performance than coherently detected QPSK.

5   Multi-amplitude and Phase Shift Keying MAPSK

Both amplitude and phase shift keying schemes have their limitations. Better results can often be obtained by a combination of both amplitude and phase shift keying. This makes more efficient use of the same transmitted power, as it requires less power than for PSK to achieve the same error probability. It also enables the number of amplitude levels to be optimised as compared with ASK. The received signal can be reconstructed in the presence of a lower signal-to-noise ratio as compared with either ASK or PSK. An example of a 16-level (2 amplitude, 8 phase) scheme is shown in Figure 9.5. A special case of MAPSK is quadrature amplitude shift keying (QASK), also known as quadrature amplitude modulation (QAM). In this scheme, each of the four transmitted dibit states are produced by quadrature combinations of the amplitudes ±1, that is the signal produced can be represented by

$$S(t) = X(t) \cos \omega_c t + Y(t) \sin \omega_c t \tag{9.3}$$

where $X(t)$ and $Y(t)$ take the value of +1 or −1 according to the corresponding data bit values. The number of states in QAM can be increased only by adding more amplitude levels to each of the quadriphase signals. For example, Figure 9.6 shows how 16 possible states can be produced from amplitudes ± 1, ± 3. Four-level QASK produces an identical transmitted output to QPSK, and has identical error performance.

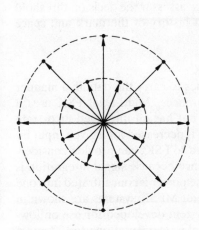

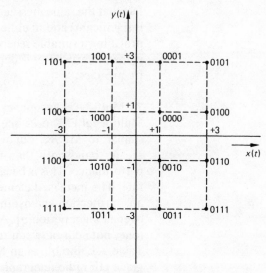

Fig. 9.5 *Polar diagram for 16-level MAPSK*   **Fig. 9.6** *Four-level QASK*

6   Frequency Shift Keying FSK

In frequency shift keying, digital information is transmitted by using, as a code, the sequential transmission of carrier pulses of constant amplitude and several different frequencies.

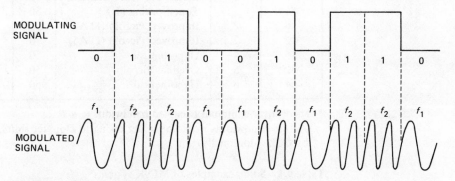

Fig. 9.7 *Frequency shift keying of a binary pulse stream*

Binary FSK uses two frequencies (see Figure 9.7) with a minimum separation usually equal to twice the bit rate. Detection is normally incoherent using filters (or their equivalent) centred on the tone frequencies. Coherent detection is possible and gives a 3 dB improvement over the incoherent scheme, but is significantly more difficult to implement. An ideal coherent FSK detector could be realised by coherent detection in matched filters of two orthogonally spaced frequencies followed by comparison of the detected outputs. Since the noise in the two filters is added, the performance is 3 dB worse than coherent PSK.

Wide deviation FSK with a law assessor is useful in counteracting narrow-band interference and frequency selective fading. The principle relies on the fact that the same message is carried by both frequencies (i.e. absence of a tone is significant) and, if either fades, it may still be possible to receive the message providing a suitable receiver is used. In the law assessor the decision threshold is made variable and dependent on the past history of the mark and space frequencies.

7   Multi-level Frequency Shift Keying MFSK

Multi-level FSK uses one of a number of tones to carry the data in a manner similar to MPSK, but with important differences. Multi-level FSK uses a greater bandwidth than the two-level case, but has an improved error rate performance. This is because the symbol rate is decreased for a given input bit rate. The increased element length, compared to FSK, can provide considerable protection against intersymbol interference because signal integration is over a longer period[3]. At any one time all the power is concentrated into one tone, not split between the tones. Examples of MFSK systems are shown in Table 9.1. Piccolo is an MFSK telegraphy system developed for use on low-grade HF radio telegraph circuits. The original implementation used 32 audio tones to represent the 32 characters of the teleprinter alphabet (Murray code).

NO. OF TONES (n)	BITS PER ELEMENT ($m = \log_2 n$)	TITLE	DATA RATE EQUIV-ALENT (bit/s)	ELEMENT DURA-TION (T msec)	OCCUPIED BAND WIDTH (Hz*)
2	1	Ideal FSK with Minimum Tone Spacing ($1/T$)	75	13.3	150
			50	20	100
6	2.5	Improved Piccolo (ITA2)	50	50	120
12	3.5	Improved Piccolo (ITA5)	70	50	240
16	4	Hypothetical	50	80	200
32	5	Original Piccolo	50	100	320
64	6	Hypothetical	50	120	533

* Theoretical minimum occupied bandwidth = n/T
(Actual bandwidths would be wider owing to Doppler shifts, etc.; that is at least ±100 Hz additionally)

Table 9.1   Some examples of MFSK systems

To be compatible with 75 baud teleprinters Piccolo was designed to operate at 10 characters per second. Thus each tone lasts one tenth of a second. The incoming serial stream is converted to a parallel format which is then used to select the transmitted tones. When there is no data to transmit, a standby tone is transmitted and is used for synchronisation purposes.

9.2.3 Phase Continuous Schemes

1 Binary FM

Binary FM is, as its name implies, the switching of a carrier frequency between one of two values, according to whether the incoming data is a 1 or a 0. Binary FM is also known as continuous phase frequency shift keying (CPFSK). The optimum deviation ratio for narrowband binary FM is 0.715. (*Deviation ratio* is defined as the ratio between peak-to-peak deviation and bit rate.) The transmitted spectrum is compact with this deviation, thus reducing adjacent channel interference (which can be controlled further by suitable shaping of the baseband signal at the transmitter). Excessive filtering at the baseband will, however, introduce unacceptable intersymbol interference. Additional bandwidth reduction may be obtained through the use of deviation ratios below the optimum value but this is at the expense of reduced power efficiency. In practice the choice of deviation ratio must be a compromise between occupied bandwidth and performance of the system in noise.

The modulation spectrum of binary FM is shown in Figure 9.8. A deviation ratio of 0.715 and no pre-modulation filtering is compared with a deviation ratio of 0.63 with pre-modulation filtering.

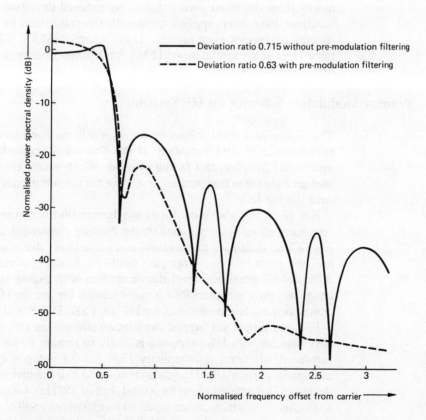

Fig. 9.8 *Modulation spectrum of binary FM*

2   Amplitude Modulated FM

Band-limiting of FM signals can produce distortions of the RF signal which introduce amplitude variations on the transmitted envelope. Conversely, the application of suitably tailored amplitude modulation on to an FM carrier can reduce the occupied bandwidth of the FM. The applied AM must be carefully derived from the data stream being transmitted. Bandwidth reduction in this way introduces a power penalty of about 1 dB. This technique has also been called *simulated filtering*.

3   Minimum Shift Keying MSK

Minimum shift keying is a special case of binary FM in which the deviation ratio equals 0.5. The spectral occupancy is lower than for optimum binary FM, where the deviation ratio is approximately 0.7. Thus more data can be transmitted, so that MSK is sometimes known as fast frequency shift keying (FFSK). Detection is usually performed coherently, although the more inefficient differentially coherent and non-coherent demodulation techniques can be employed. When phase information is gathered over two adjacent bit periods, MSK is more power efficient than conventional non-coherent FM. However, conventional binary FM (with deviation ratio = 0.7), when detected coherently and with a three-bit observational interval, is theoretically more power efficient than MSK. Thus the advantages claimed for MSK arise primarily from the band saving due to its reduced deviation. Various shaping functions have been applied to smooth the transitions in MSK and reduce spectral occupancy even further. These schemes (SFSK, GMSK), along with tamed frequency modulation (TFM), have power efficiencies of the order of 1 dB worse than MSK[4].

9.2.4   Primary Modulation Schemes for HF Systems

The choice of a good robust modulation scheme is critical to the successful performance of an HF system. If the channel is unsatisfactory because of multipath, Doppler, fast fading or other effects which are not noise related, then an increase in transmitter power will not in itself produce an improvement over the HF link.

Not all of the above modulation schemes find common usage within HF communications systems. Rather, the limiting constraints of frequency selective fading, multipath propagation and noise imply that a number of schemes are not practical. For example, multi-amplitude systems suffer from the difficulty of amplitude level discrimination with fading signals or on noisy channels; they are therefore a poor choice for use in HF systems. Three common modulation methods for HF are FSK, DPSK and MFSK.

FSK is a simple yet rugged modulation scheme for HF; for this reason it is very popular with HF users and is likely to remain so for some time in the future. Apart from its simplicity FSK has advantages in its immunity to multipath, particularly if the delay is equal to half the inverse of the frequency deviation (for example 2 ms for a total shift of 250 Hz). Under these conditions a selective fade which causes a null on one frequency will cause a maximum on the other, and since the demodulation process in modern FSK systems can operate on either tone independently, this provides an effective diversity operation. However, for multipath delays of about 4 ms or more severe timing errors are introduced and the system fails rapidly. A standard frequency separation of 850 Hz between tone frequencies is often adopted.

DPSK modulation can track slowly varying characteristics of the ionosphere because the system uses the received phase of the previous element as phase reference. It is therefore able to 'adapt' to changing conditions on the link. Where the highest possible information density is required a wideband multi-tone DPSK system (section 9.4.2) has a definite advantage provided that the link has an adequately high signal-to-noise ratio and that extreme ionospheric conditions are not encountered. Fast fading can be a problem with DPSK and the use of a guard band against multipath is essential. Unfortunately the provision of a guard band causes an effective power loss in the system.

The superiority of MFSK is most evident against the timing error effect of multipath, which is acheived by use of a long element length. For MFSK, error rates rise slowly as the multipath delay increases and can be quite acceptable even for very long delays. Other modulation schemes fail rapidly once the delay exceeds a small fraction of the element length. MFSK is also relatively immune to Gaussian and impulsive noise and in wideband interference. Its main disadvantage lies in the higher demands made on the stability of the radio system and the increased implementation complexity.

9.3 Error Rate Performance

9.3.1 Error Probability

A performance comparison of the different modulation types is made most easily in terms of error probability. In previous chapters an assessment of system performance was made in terms of the prevailing signal-to-noise ratio on the channel. For a given modulation and detection method the signal-to-noise ratio is directly related to the probability of *element error* (each symbol to be conveyed in the message is made up of an integral number of *elements* – element error is synonymous with bit error if one binary element conveys one bit of information).

Let P_e equal the probability of element error in the message. If

$$R = E_b/N_0 = \frac{\text{signal energy per element}}{\text{noise per unit bandwidth}}$$

then the signal-to-noise ratio, S/N, used in the usual sense of signal power divided by noise power in the bandwidth in use is given by

$$\text{S/N} = (E_b/N_0) \times \frac{\text{bits per second}}{\text{bandwidth in hertz}} \tag{9.4}$$

The error characteristics of the common modulation schemes are now discussed; the reader is referred to the book by Ralphs[3] for a more detailed examination of MFSK.

9.3.2   Non-fading Signals

The relationship between error probability and E_b/N_0 for digital receiving systems is first considered with reference to ideal receivers, receiving non-fading signals in the presence of Gaussian noise.

Equations[5] for the non-fading condition fall into two distinct categories:

1   *Coherent detection*

The coherent systems are each represented by an error function curve:

$$P_e = \tfrac{1}{2}\,\mathrm{erfc}\,(aR)^{\frac{1}{2}} \tag{9.5}$$

where $a = 1$ for PSK
$a = \tfrac{1}{2}$ for FSK and ASK
$R = E_b/N_0$

2   *Non-coherent detection*

The error rate of each of the binary non-coherent systems is represented by an exponential curve, so that the difference in performance between any two of them can be expressed in a single figure representing the change of signal-to-noise in decibels, required to maintain a constant character error rate. The expression is

$$P_e = \tfrac{1}{2}\exp\,(-bR) \tag{9.6}$$

where $b = 1$ for DPSK

$b = \tfrac{1}{2}$ for ASK and FSK (matched filter reception and envelope detection).

Figure 9.9 shows the performance of various methods of modulation and reception. The following points are relevant to the interpretation of the curves:

a) The error probability is in terms of message bits.

b) The signal energy per message bit is the average of the mark and space signals. This is important when considering amplitude modulated signals, because signal energy is sometimes defined as the energy in the 'carrier on' state.

c) For the quaternary systems the energy per signalling element is twice the energy per message bit. This has been taken into account in constructing the curves.

d) The systems are assumed to be 'ideal' and the pre-detector bandwidth is the minimum needed theoretically. This means that for a data rate x, in message bits per second, then the minimum bandwidth is x Hz for BPSK, binary FSK and ASK; it is $\tfrac{1}{2}x$ Hz for QPSK.

9.3.3   Fading Signals

HF signals received over sky wave paths exhibit variations in signal strength with time. As described in Chapter 4, the HF medium is concerned with short-term fading, which is periodic, with fading periods in the range 10 ms to 100 s.

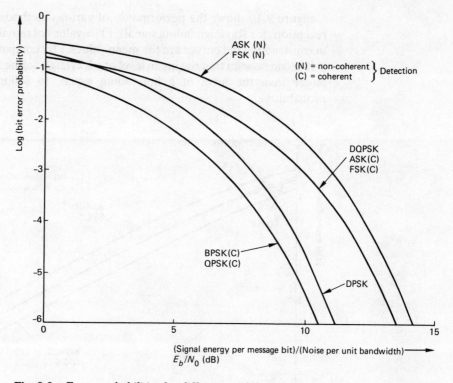

Fig. 9.9 *Error probabilities for different modulation schemes under non-fading conditions*

The amplitude distribution of the signal caused by short-term fading can be regarded as following the Rayleigh distribution.

The method of calculation[5] involves the assumption that the change of signal level is small within the duration of one signalling element. This will be the case for the data rates of interest (i.e. 50 bit/s to 2.4 kbit/s), except for the fastest fading rates, i.e. so-called 'flutter fading'.

Under conditions of slow flat Rayleigh fading the equivalent expressions for probability P_e of element error become:

a) *Coherent detection*

$$P_e = \tfrac{1}{2}[1 - (1 + 1/aR)^{-\frac{1}{2}}] \tag{9.7}$$

$$\simeq 1/(4aR) \text{ for large } R$$

b) *Non-coherent detection*

$$P_e = \tfrac{1}{2}(1 + bR)^{-1} \tag{9.8}$$

$$\simeq 1/(2bR) \text{ for large } R$$

where *a*, *b* and *R* have the same meaning as before.

Figure 9.10 shows the performance of various methods of modulation and reception, for Rayleigh fading signals. (The values of signal energy per bit used in constructing the curves are the mean values.) By comparing Figures 9.9 and 9.10 it can be seen that the mean level of a Rayleigh fading signal must be much larger than the level of a non-fading signal, to achieve the same error probability.

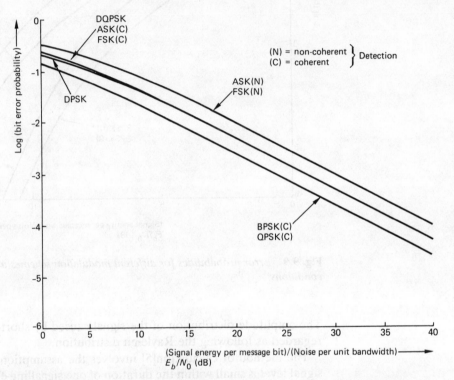

Fig. 9.10 *Error probabilities for different modulation schemes under slow flat Rayleigh fading conditions*

9.3.4   Use of Diversity Reception

The effects of fading can be reduced by receiving, and combining, more than one version of the signal (the fading of the different versions being uncorrelated). The effects[5] of diversity operation on error probability can be calculated as follows:

a)   Coherent detection

$$P_e = \tfrac{1}{2}(1 - Z^{1/2}) \tag{9.9}$$

$$\simeq 3/(16a^2R^2) \text{ for large } R$$

where $Z = (1 + 3/2aR)^2/(1 + 1/aR)^3$ for dual diversity.

b) Non-coherent detection

$$P_e = \tfrac{1}{2}(1 + bR)^{-n} \tag{9.10}$$

$$\simeq \tfrac{1}{2}(bR)^{-n} \text{ for large } R$$

for *n* path diversity.

Figures 9.11 and 9.12 show error probability versus E_b/N_0 for various types of modulation and reception, for single and dual diversity paths. It is assumed that the mean E_b/N_0 is the same in all diversity paths, and that the correlation between the fading in the different paths is zero. The path combination is assumed to be performed by maximal ratio combining, i.e. by combining the paths after weighting in accordance with the signal energy in the individual paths.

The following types of diversity reception are possible:

a) Spaced-antenna diversity.
b) Polarisation diversity.
c) Elevation angle of arrival diversity.
d) Frequency diversity.
e) Time diversity.

Types *a*, *b* and *c* all relate to antennas, and have already been discussed in Chapter 6. Type *e*, time diversity, is related to coding, and is discussed in section 9.6. Type *d*, frequency diversity, is discussed below.

Diversity operation can be obtained by transmitting the same information at different frequencies. If frequency spacings of many kilohertz are used, zero correlation of short-term fading can be guaranteed; in practice this method of diversity is rarely used because of the need for spectrum conservation. However, levels of correlation[6] below unity can be obtained at frequency spacings down to about 400 Hz. This is *selective fading*. At these narrow spacings the decorrelation between frequencies is primarily a function of the sky wave mode structure; in the presence of two major modes the optimum frequency spacing in Hertz is half the inverse of the path time difference in seconds[7].

In current practice, frequency diversity with narrow spacings is met in two forms:

1 The two tones of a binary FSK transmission are treated at the receiver as two on-off keying (OOK) signals carrying the same information, the tones being isolated by filters and demodulated by envelope detectors. The demodulator outputs are combined using the so-called 'assessor' circuit. The ideal performance of such a system can be obtained using the two-path curve of Figure 9.12; however, it must be remembered that the power in each diversity branch is only half of the total power.
2 Data is conveyed on a number of sub-channels, which are modulated using narrowband FSK, or binary or quaternary differentially coherent phase modulation. This technique is the principle used in multi-tone modems. When the full data capacity represented by all the sub-channels is

not needed, non-adjacent sub-channels are grouped to give narrowband frequency diversity. An example of this approach is Kineplex, as discussed in section 9.4.2.

Diversity reception, and methods of path combination, are discussed more fully by the CCIR[8].

9.3.5   Non-ideal Receivers

The performance of practical receiving systems falls short of that of the ideal systems discussed above. There are many reasons for this, but the most important are non-ideal filtering, non-optimum threshold setting, and errors in sampling time.

The short-fall in performance of any particular receiving system can be related to the performance of the equivalent ideal system by including[5] a demodulation factor. The *demodulation factor* is the number of decibels by which the signal-to-noise ratio of the practical system must be increased above that of the ideal system, for the same error probability to be obtained. The demodulation factor may vary with error probability, but for most systems the demodulation factor can be regarded as constant over the range of error probabilities of practical interest.

It is very difficult to measure the demodulation factor of a receiving system with real signals because the propagation situation is not repeatable; it is, however, possible to compare two or more systems by having them operate in parallel. The more usual procedure is to make measurements using a fading simulator. Practical receiving systems typically have demodulation factors of around 4 to 5 dB.

9.3.6   Effect of Non-Gaussian Noise

The discussion so far has assumed that the noise present is Gaussian. This is normally true for VHF and UHF systems, which are internally noise limited. At HF external noise limitation usually prevails, except at the high frequency end of the band, at very quiet sites, and in circumstances which necessitate the use of inefficient antennas.

In the presence of impulsive noise the error rate is much less sensitive to signal level than is the case with Gaussian noise[9]. All the error probability versus E_b/N_0 curves presented so far would thus have flatter slopes if the noise contained an impulsive component. This effect is very difficult to quantify, but must be taken into account when considering the appropriate methods of error control coding.

9.3.7   Multipath and Interference Considerations

Interference from a CW signal does not affect all modulation schemes equally but can cause serious degradation in differential and multi-level schemes. An error in one bit will then produce an error in the following bit.

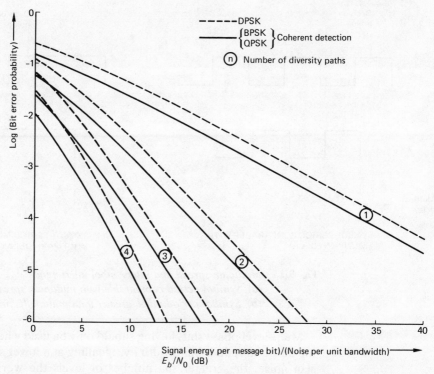

Fig. 9.11 *Error probabilities for PSK modulation schemes with diversity reception*

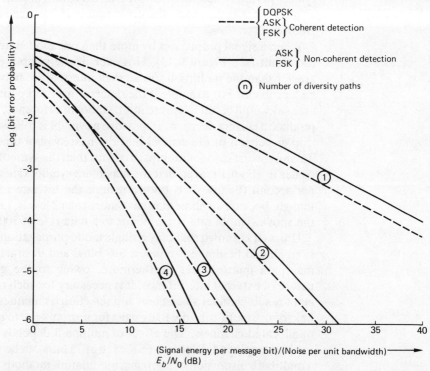

Fig. 9.12 *Error probabilities for ASK and FSK modulation schemes with diversity reception*

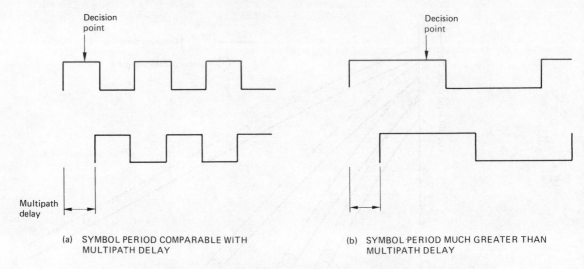

(a)   SYMBOL PERIOD COMPARABLE WITH
        MULTIPATH DELAY

(b)   SYMBOL PERIOD MUCH GREATER THAN
        MULTIPATH DELAY

Fig. 9.13   *Multipath spread and intersymbol interference*
(a) Symbol period comparable with multipath spread
(b) Symbol period much greater than multipath spread

Multi-level phase shift keying should only be used when there is an overriding desire to reduce bandwidth by signalling at a lower symbol rate. In Gaussian noise, the greater the number of levels the worse the bit error rate performance. This may not be the case for a multipath channel because the symbol period is increased. QPSK is an exception in that, because the two pairs of phases are orthogonal, the error rate performance is the same as that for BPSK.

When a signal propagates by more than one mode intersymbol interference can result (see Figure 9.13). However, if the symbol period is significantly greater than the multipath spread then interference between symbols will not be a problem. For example if the detector decision point were as shown then case (a) would result in severe intersymbol interference, whilst case (b) would produce a correct decision. In practice a symbol is followed by a guard band to stop dispersion of the symbol into the next symbol time slot at the receiver. Taking a worst case for multipath of 8 ms then the symbol plus guard band time comes to about 10 ms so that the maximum symbol rate would be 100 symbols per second (baud). This does not limit the bit rate to 100 bits per second though, because symbols may have more than 2 levels, i.e. an 8-level symbol at the above symbol rate corresponds to a data rate of 300 bits per second.

Unless it is known that only a single mode propagation exists, FSK transmissions should be limited to below 300 bit/s, and in practice much less, due to multipath interference. Furthermore, owing to the generally narrowband nature of external interference, it is necessary for such transmissions to either employ sub-channel assessment and selection techniques, requiring a two-way link, or use multi-tone diversity pairs for transmission to make the optimum use of allocated channels. The effect of multipath depends upon the modulation method employed. Whereas FSK even at 75 bit/s can be very badly affected by a multipath dispersion of say 6 ms, modulation methods such as MFSK will be impervious up to some 16 ms.

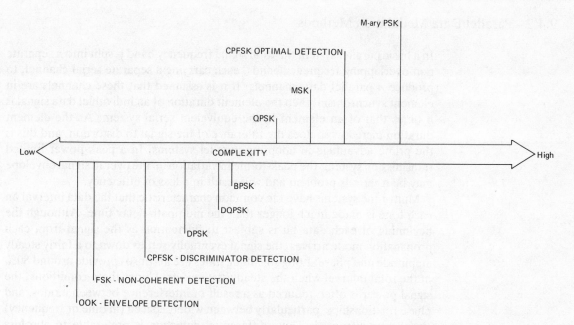

Fig. 9.14 *Relative complexity of various modulation schemes*

Figure 9.14 gives an idea of the relative complexity of the different forms of modulation, and hence expense to implement. Coherent demodulation is preferred and, in poor conditions especially, several dBs advantage over non-coherent demodulation may be obtained. Differentially coherent demodulation performs nearly as well as full coherent demodulation, though differentially encoded schemes are seriously degraded in an interference environment. For higher rate transmissions, in the range 1200 – 2400 bit/s, the time dispersion due to multipath propagation may necessitate multiplex techniques involving parallel sub-channels.

9.4 High Data Rate Transmissions

9.4.1 Types of Technique

The fundamental problem involved in increasing the data rate of HF transmissions is that of intersymbol interference caused by multipath effects. Two techniques which are available to overcome these problems are

a) Parallel data transmission methods, in which the high speed data is sent as a number of independent low data rate channels. The signalling rate must be kept well below the reciprocal of the multipath delay to avoid the effects of intersymbol interference. A higher data rate can be effected by multiplexing the data stream within a number of different tone frequencies.

b) Equalisation techniques, which attempt to correct distortions caused by the channel.

9.4.2   Parallel Data Modulation Methods

In a basic parallel system the total signal frequency band is split into n separate non-overlapping frequency bands, each carrying a separate serial channel, to produce n parallel data channels. If it is assumed that these channels are in element synchronism, then the element duration of an individual data signal is n times that of an element in the equivalent serial system. As the element duration increases so does the tolerance of the signal to distortion, and this is the prime advantage in adopting parallel systems. In a peak-power limited transmission system, the peak-to-mean variation on the overall signal envelope may be a serious problem and will result in a loss of efficiency.

Multi-tone systems have the common characteristic that the data interval on each tone is made much longer than the multipath delay time. Although the beginning of each data bit is subject to fluctuation as the signal from each propagation mode arrives, the signal eventually settles down to a fairly steady amplitude and phase. Long signalling times (12 ms or so) provide around 80% of the total interval when the steady state exists. Under these conditions, the signal power is often reduced as a result of interference between modes, and phase relationships, particularly between widely spaced (in time or frequency) signals are disrupted. Thus, differential detection is preferable to absolute phase modulation and a more up-to-date comparison can be obtained if it is made simultaneously between adjacent tones.

If an efficient use of bandwidth is not a priority, the most effective parallel system occurs when the spectra of the individual channels do not overlap and conventional filters can be used to isolate each individual channel. This represents an inefficient use of bandwidth because of the frequency guard bands that are between the tones. The idea of using overlapping spectra has therefore received attention as this means that the total symbol rate over a bandwidth of B Hz can now approach $2B$ symbols per second, which is the ideal Nyquist rate[10].

There have been a number of different approaches to improving the data rate on the HF channel using differential phase shift keying (DPSK):

1   Time DPSK

Probably the first practical implementation of DPSK was the Kineplex[11] equipment.

Kineplex is a high-capacity data transmission system that can operate in a single telephone voice band or equivalent. The system accepts binary input and generates many parallel data channels. Kineplex is a four-phase multi-tone time DPSK system in which the information is contained in the relative phase difference between sequential bits transmitted on the same tone. The phase of the received bit is stored, and compared with the phase of the next bit. Variations in the channel response which are slow compared with the bit duration give similar distortion to bits adjacent in time, so that the relative phase difference is appoximately unchanged. One Kineplex version uses 20 tones, with each tone composed of two independent sub-channels. Transmission occurs with 75 baud per sub-channel, providing a total capacity of 3000 bit/s.

The elements in the different channels are in element synchronism, and time is set aside for guard bands to provide a tolerance to multipath. Two binary phase modulated signals of the same frequency but in phase quadrature are combined to produce a four-level phase modulated signal. The carriers are spaced in frequency by the reciprocal of the element detection period, which itself is less than the element duration. The full 20-tone system occupies a bandwidth of 3.4 kHz and this includes 500 Hz frequency guard bands at the band extremities. The synchronizing tone is at 2915 Hz if it is used, if not then stable timing generators must be employed. The synchronisation is to keep the detector in step with the incoming signal pulses. Differentially coherent detection is employed by the receiver which uses a pair of filters to work on the differentially encoded QPSK signal. The latter detection technique is used to avoid problems involved in extracting reference carriers for coherent detection.

2 Interpolated PSK

This technique has been developed to overcome some of the shortcomings in the Kineplex system. In a fading channel de-correlation of the phase reference from symbol to symbol results in the modem having an irreducible bit error rate (BER), regardless of signal-to-noise ratio over the link. For example, a 16-tone Kineplex has a BER of about 10^{-3} at 2400 bit/s throughput. One form of improved multi-tone modem employs a tone library of 39 tones and uses time differential QPSK on each channel. The tone spacing is the reciprocal of the integration interval over which coherent detection of the symbols take place. Interpolated PSK, or IPSK, attempts to produce a phase reference for data detection. Alternate tones are modulated either with a reference phase or with data. At the receiver the reference phase information is extracted from the reference tones and an algorithm is used to obtain phase information for tones between reference tones. Interpolation both in time and in frequency are used to generate the local references. This is computationally fairly intensive and has been implemented with a digital signal processor chip and associated hardware.

3 Frequency DPSK

In frequency DPSK systems, the information is coded as the phase difference between two tones transmitted simultaneously on different frequencies. Differential comparison is thus performed under multipath conditions which are closely correlated. A better performance might be expected than is achieved for time DPSK systems at the expense of additional equipment complexity. The individual tone oscillators must be coherent and the starting phase of each must be known.

In the absence of multipath, this system is identical to the time DPSK system discussed above and has an error performance which is 3 dB worse than conventional PSK. Frequency DPSK systems require additional equipment complexity to produce carrier synchronisations, but they can be expected to perform rather better than time DPSK systems when used over time dispersive channels. This is because the differential comparison is carried out under the same multipath conditions for each tone. Thus, the correlation operation that is performed in the phase comparators is more nearly ideal.

Andeft[12] is a four-phase frequency DPSK system which uses 40 Hz tone separation, with 66 tones transmitted simultaneously, two of them being used for synchronisation and as starting reference frequencies. Each tone is compared with the adjacent tone to detect the relative phase. Whole frames are detected simultaneously using 64 sets of correlators, modulators and phase detectors.

9.4.3   Equalisation Techniques

The multi-tone or parallel HF modem has, for some years, provided the basic means of data communication at high rates. Adaptive serial techniques now represent a step forward in the realisation of the potential for HF.

These techniques usually involve the application of some form of correlator at the receiver in order to 'equalise' effects produced by the channel. The *channel transfer function* can be described in terms of

a) an impulse response in the time domain, or
b) a frequency response in the frequency domain.

There are two essentially different strategies for *a*. One approach attempts to combine all multipath components, whilst the other cross-correlates a single mode in an effort to reject all others. The approach for *b* uses a filter which produces a flat frequency response and a linear phase response (i.e. a filter with a frequency transfer function which is the reciprocal complex conjugate of the channel).

For either *a* or *b*, the equalisation networks gradually become out of date because of the time-varying nature of the channel. Thus, any adaptive system operating over an HF channel must be able to

(i) measure the channel parameters (which may involve transmitting a probe signal known to the receiver);
(ii) set matched filter elements at the receiver according to the measured parameters;
(iii) repeat (i) and (ii) sufficiently often to follow variation of the channel.

The equaliser, which is basically a digital filter, must store samples which span the expected multipath spread and multiply each sample by a coefficient before summing to produce the symbol to be decoded. This can be visualised as a tapped delay line structure, with the symbols shifted along the line as each new one arrives. Once the coefficients have been determined, this structure can overcome the multipath problem but still has to cope with the varying channel.

A measure of how much the tap coefficients are in error is needed and a least mean square error criterion has been found to be the most effective. The transmitted data sequence and the estimated received sequence are compared to produce a measure of the error. With simple binary signalling there is enough information to train an equaliser from start-up just relying on equal probability of symbols, but multi-level signals would need a training sequence which implies an increase in redundancy of the message.

Two early attempts at systems incorporating adaptive equalisation were Adapticom and Kathryn.

The Adapticom[13] system applies time domain adaptation to the reception of a 2660 bit/s serial transmission for which the bit duration (about 0.4 ms) is considerably less than the multipath time spread; the spectrum of each bit occupies the whole voice bandwidth. The matched filter operates on the baseband and its response is determined by tap amplifier gains. A single transmitted probe pulse is received as a sequence of multipath signals. Instantaneous values of the composite received signal are sampled at the delay line and stored as tap amplifier gains. The matched filter equalises the delays, squares all amplitudes of multipath spectral components and adds them coherently, thus compressing each signal into a duration comparable to that for the transmitted signal. The matched filter characteristics in the receiver must be updated frequently to follow changes in the channel transfer function.

The Kathryn[14] system applies frequency domain adaptation to the reception of a PSK multi-tone signal. An overall data rate of 2550 bit/s is achieved using 34 orthogonal tones transmitted in 75 baud frames, each frame carrying information and probe signals in phase quadrature. A 1 ms guard time is used to attempt to maintain orthogonality between tones under multipath conditions. The receiving terminal uses two receivers for diversity operation with optimum combining. A pseudo-random probe sequence, identical to that at the transmitter, is generated in synchronism at the receiver. The measurement section of the receiver examines each received tone and determines the amplitude and phase distortion of the probe component. The latter is then removed from each tone and the remaining information component corrected in phase and modified in amplitude by the weighting section. The measurement and information weighting sections of the receiver form the adaptive matched filter. Continuous updating is provided using the probe sequence. However, the system possesses some inertia so that the receiver phase reference can never precisely compensate for the current state of the channel.

9.4.4 Comparison of Techniques

Serial data transmission with pulses narrow enough to resolve multipath can achieve[15], with optimum equalisation, an irreducible bit error rate (BER) that is a number of orders of magnitude less than the irreducible BER of parallel data transmission using multi-tone systems.

The applicability of adaptive matched filter techniques relies on the channel multipath impulse response remaining constant for significantly longer than the duration of the echo train from a single sample. Thus, the product X of the multipath spread and Doppler spread must be appreciably less than unity for adaptive systems to be useful. The transmission limit of communication for parallel data modems occurs[15] for $X \approx 1/2000$, whilst for serial data and adaptive equalisation, it occurs for $X \approx 1/200$. In practice, however, the performance of an adaptive system has been worse than the performance of non-adaptive multi-tone systems. The major problem was undoubtedly caused by the rapid variability of the channel, which makes the greatest demands on equaliser performance when the signal is at its poorest. The first systems were developed in the late 1960s when the necessary degree of processing power was not readily available. The availability of high speed signal processing hardware now permits real time adaptation with sufficient capability to track the fast changing HF channel. The characteristics of the older systems described above are summarised[16] in Table 9.2. Of attempts such as Kathryn, Andeft, Adapticom and Kineplex, only Kineplex has entered service on point-to-point circuits.

SYSTEM NAME	PROCESS-ING DOMAIN	DATA RATE (bit/s)	MODU-LATION	ADAPTIVE PROCESSING
Adapticom	time	2660	DPSK	Matched filter.
Kathryn	frequency	2550	PSK (34 orthogonal tones)	Diversity - local reference correlation.
Kineplex	time	3000	DPSK 4-phase (20 tones)	Relative phase comparison on same tone.
Andeft	frequency	4800	DPSK 4-phase (66 tones)	Relative phase between two frequencies.

Table 9.2   Summary of characteristics for some high data rate modulation schemes

The peak-to-mean variation of overall signal envelope power for parallel data transmission systems can be a serious problem, resulting in a loss of efficiency.

The possibility of an in-phase addition of the tones results in a large peak-to-mean ratio of the transmitted signal; peak limiting or clipping may be used to limit this to about 6 dB. In comparison the 3 dB peak-to-mean ratio of a serial modem represents a halving of the transmitter power requirements.

Frequency selective fading or narrowband interference can notch out one or more tones from a parallel modem's tone set, thus resulting in data errors. The serial modem is able to exploit the implicit signalling diversity provided by multipath, assuming that it has adapted correctly to the channel. In the presence of fast fading this may not always be possible, however.

Doppler shifts become important when parallel schemes with overlapping spectra are used. In this case frequency shifts will result in inter-channel interference.

The major problems with the serial approach are tracking the rapid changes in the HF channel, and the implementation complexity. The computational load of a serial modem is far greater than for a parallel modem despite the latter having to handle many channels. Algorithms for adaptive systems must produce a channel model very rapidly and it is this rather than the compensation for the channel that is difficult computationally.

Given optimal equalisation and detection a serial modem will have a fundamentally lower irreducible bit error rate than a parallel modem. However, there is still some work to be done in developing the adaptation algorithms such that the equaliser can efficiently track the variations of the HF channel. Quite significant processing power is needed, and the basic data rate or message length must be increased to add training sequences. The serial approach is likely to dominate eventually, but the rate of technical progress will dictate when it will actually happen.

There would be significant advantage to be gained in equipment commonality and flexibility if the same modulation scheme could be used for all types of messages. Parallel schemes such as Kineplex rely on the tone spacing being the reciprocal of the symbol detection period to maintain orthogonal channels, so the data rate cannot be efficiently reduced. Serial schemes thus seem to be more promising if commonality is a system goal. However, at present the system designer may be justified in opting for the established, proven techniques of the parallel tone solution. Until a clear standard emerges for serial modems, their acceptance within HF communication systems may be slow.

9.5   Wide Bandwidth Techniques

9.5.1   General Limitations

Systems that employ data transmissions in the HF band have generally been designed for operation with voice channel equipment, and signal bandwidths have so far not exceeded a few kilohertz. At frequencies reasonably close to the optimum working frequency such bandwidths have provided useful protection against selective fading. When considering wideband modulation methods for HF sky wave transmission, it is relevant to examine the bandwidth limitations of the channel. The absolute maximum channel bandwidth, based purely upon signal-to-noise ratio, ranges from the LUF to the MUF.

The LUF is determined by the minimum acceptable signal-to-noise ratio. Frequencies below the LUF provide ratios that are too poor for useful communication. The LUF depends upon the radio equipment characteristics such as transmitter power and antenna gain and upon external noise and interference levels.

The MUF, however, is purely a characteristic of the state of the propagation path through the ionosphere; it is independent of radio equipment and noise level. The maximum bandwidth therefore depends upon the maximum accept-

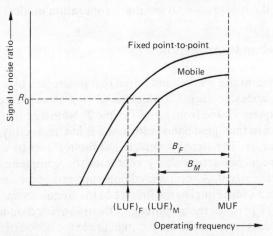

Fig. 9.15 *Maximum operating bandwidth*

able error rate determined from the just tolerable signal-to-noise ratio R_0. This bandwidth B_F for a fixed ground station might range from a few megahertz during favourable daytime conditions to less than 1 MHz at night. B_F is shown schematically in Figure 9.15. An HF mobile would, by contrast, have a higher LUF and consequently a smaller available bandwidth B_M, because of a much smaller effective radiated power as a result of reduced antenna efficiency and transmitter output power.

The maximum bandwidth, the difference between the MUF and the LUF, is in practice not achieved; the bandwidth is usually limited by channel dispersion and interference.

9.5.2   Channel Dispersion Effects

The most convenient way of determining the echo structure of signals transmitted obliquely is by the ionosonde technique, but with the transmitter and receiver at different ends of the path. These oblique incidence sounders as described in section 8.3, use pulses of 10 µs to 100 µs duration. They have provided the direct information on multipath characteristics of ionospheric paths that is required for the design of wideband communication systems. Two examples are shown in Figure 4.21. Evidence from HF over-the-horizon radar shows that range resolution requires a wide signal bandwidth, but that it is seldom the ionosphere can effectively support an instantaneous bandwidth greater than about 100 kHz. The problems that arise from delay distortion differ according to the type of wideband signal in use. For a true spread spectrum signal, the problem of acquiring and maintaining synchronisation to a fraction of the spreading sequence period is extremely difficult, even in the absence of multipath. For spread bandwidths of 500 kHz or greater, an accuracy of synchronisation significantly less than 1µs would be required, which implies an improvement of several orders over existing HF modems. The synchronisation problem becomes more acute if the dominant propagation mode changes across the signal bandwidth; this feature can act as a significant bandwidth limitation. A frequency hopping system, however, may only require bit synchronisation at the data keying rate, although care must again be taken if the hop crosses to another propagation mode.

9.5.3   Matched Filter/Wideband Correlation

Some systems use a wideband signal that is detected by a matched filter. As an example consider chirp[17] which consists of a swept frequency signal traversing a frequency band from f_0 to f_1 in time T. Note that $(f_1 - f_0)$ is not necessarily restricted to the signal bandwidth and T is not necessarily the same as the input data interval. The signal is detected at the receiver by a matched filter which has a frequency-group delay characteristic complementary to that of the transmitted signal. A compressed pulse appears at the output of the filter. This pulse has a width equal to the inverse of the frequency spread and an amplitude $[(f_1 - f_0)T]^{\frac{1}{2}}$ times the amplitude of the transmitted signal. The key to achieving high data rates with a chirp system lies in the action of the matched filter and the time compression that it produces.

If a filter is matched to a linear frequency sweep, the output pulse envelope is of the form $(\sin x)/x$. Shaping of the transmitted signal envelope, and/or the use of a non-linear frequency sweep can also be used to suppress the pulse sidelobes and to concentrate the signal energy into the main pulse. If a sequence of swept signals (see Figure 9.16) is transmitted, each of duration T seconds at intervals $\Delta t = 1/(f_1 - f_0)$ seconds apart, it is therefore possible to produce a sequence of non-overlapping compressed pulses at the output of the matched filter in the receiver. The behaviour over an HF channel is then as follows:

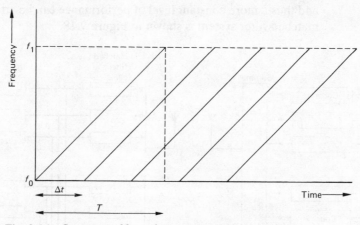

Fig. 9.16 *Sequence of linearly swept overlapping signals*

a) Under moderate multipath conditions, the chirp system performs better than conventional multi-tone DPSK systems because the differential phase comparison is effectively made between signal components separated by the signal interval Δt, rather than the longer signalling interval T appropriate to multi-tone systems. Thus, slow changes in the multipath conditions can be more effectively followed.

b) Under extreme multipath conditions, there is a severe problem of inter-symbol interference because the signalling rate is comparable with the multipath delays.

9.5.4 Matched Filter/Signal Gating

Consider a pseudo-random noise (PRN) system with overall symbol duration T, signalling rate D and individual chip width t_w (see Figure 9.17); each symbol

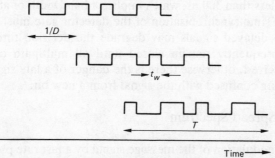

Fig. 9.17 *Sequence of pseudo-random noise signals*

is composed of a PRN sequence. The matched filter must, in this instance, pass all of the frequency components of each chip. The system bandwidth which is open to interference is therefore the reciprocal of the chip duration rather than the input signal bandwidth. There is consequently a higher probability of encountering an interfering signal and, in principle, any gains should be cancelled out on the average. In practice, however, interference is not uniformly distributed; the effect is to reduce the peaks and to fill in the troughs of the performance curve. The performance now depends upon the mean interfering power over the whole band. It is less affected by its statistical spread and thus a more constant level of performance can be attained. A generalised matched filter system is shown in Figure 9.18.

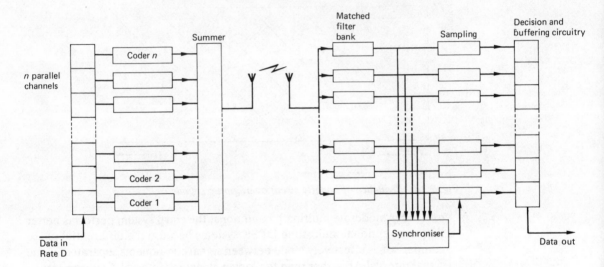

Fig. 9.18 *Block diagram of a generalised matched filter system*

It is possible to select only a single mode and to discriminate against the others by time gating. There are two major implications:

a) The output pulse width of the matched filter must be much shorter than the typical delay time between modes (i.e. the signal bandwidth must be relatively great). With typical delays of 1 ms, the pulse width should ideally be less than 300 μs, which implies a bandwidth of at least 3 kHz.
b) Time synchronisation of the detector gate must be accurate, otherwise the delayed signals may degrade the decision threshold. The gate must consequently remain closed until all multipath components have been received, otherwise there is the danger of a late signal from a previous bit being confused with the signal from a new bit.

9.5.5   Direct Sequence Spread Spectrum

The modulation of the message signal by a fast rate pseudo-random sequence can be used to produce a spread spectrum signal[18]. This provides an alternative, but mathematically equivalent, approach to the matched filter pulse compression type of system.

At the receiver, a local replica of the pseudo-random sequence is generated in synchronism with the incoming signal and is used to despread it, thus recovering the message. The system processing gain is therefore the ratio of the code chip period to the data period, the same result as derived previously. The system bandwidth which is susceptible to interference is again increased, as for the matched filter pulse compression system, and the same comments apply.

The chip length must again be much shorter than the delay time between modes. Under these conditions, code synchronisation permits the selection of an individual propagation mode. Discrimination against other modes operates with the same processing gain as for interference (see section 10.4.4). Moreover, because of the spreading action of the code, the other modes become randomised and noiselike; they do not produce the same degree of intersymbol interference. However, the spread spectrum system is less efficient in this respect than the matched filter technique, for which precise gating can completely suppress all multipath interference.

9.5.6   Narrowband Interference

The effect of narrowband interference on a multi-tone signal is to produce a high error rate on tones adjacent to the interfering signal but, with careful design of the automatic gain control (AGC), little effect on other tones. Interference nulling techniques are not effective because signal and interference, being both narrowband, are removed together resulting in no net improvement in performance. For a small number of narrowband interfering signals, the error rate is constrained without additional measures, to an upper limit no matter what the level of interference. Thus, for example, if 3 out of 32 tones are blocked the resultant bit error rate cannot be worse than 5%. However, the signal structure precludes the use of various techniques which, with other systems, might have yielded much better performance.

The processing gain of a matched filter system demonstrates the extent to which the matched filter can enhance the detectability of the signal above the noise. Consider a system whose symbol duration is T, data rate is D and overall chip width is t_w (see Figure 9.17). The transmitter codes n input data bits of duration $1/D$ into n symbols, each of duration T. These are transmitted in turn at intervals of $1/D$, with the result that if T is greater than $1/D$, then the symbols are overlapped.

At the receiver there is, for each symbol, a matched filter which produces an output pulse of duration t_w. Synchronising, gating and decision circuits (see Figure 9.18) collectively sample the matched filter outputs and assemble the decoded data into a serial output stream. The number of symbols in the channel at any one time is DT (provided $T > 1/D$). Thus relative to direct signalling, the amplitude of each symbol is reduced by $(1/DT)^{1/2}$, since the total available output power is fixed. Now, the power gain of the matched filters is T/t_w and so the overall power gain of the output pulse is

$$G = 1/Dt_w \tag{9.11}$$

However, the input signal bandwidth must be increased to produce a narrow pulse. Compared to direct rate signalling, the input bandwidth and hence the input noise power must be increased in the ratio $1/Dt_w$. Hence, the overall processing gain of the system in noise is unity and the matched filter approach has the same performance as direct PSK modulation in Gaussian noise.

However, for an individual interfering source, the dispreading process on receive spreads out the interference and, on the average, its level is reduced by the ratio of the signal bandwidth to the spread bandwidth, i.e. by the ratio $1/Dt_w$. For the chirp matched filter system the corresponding ratio is $(f_1 - f_0)/D$.

9.6   Principles of Error Control Coding

9.6.1   Causes of Data Errors

In the preceding calculation of error probability versus E_b/N_0 the distribution of errors with time was not considered. This must now be taken into account because it affects the subsequent discussion of coding. Errors can be produced by symbol corruption due to multipath time dispersion. Digital signals received over HF sky wave paths are subject to both burst and random errors. The bursts of errors tend to occur at the minima of the fading pattern.

For constant level signals in Gaussian noise the distribution of errors is random. There is an exception in the case of differentially demodulated phase shift keying, in which there is a tendency for errors to occur in pairs; however in Gaussian noise the proportion of paired errors is generally not large; it is much larger with impulsive noise. With Rayleigh fading signals in Gaussian noise most of the errors occur in bursts, at the time of minimum signal level. Random errors are still present in addition to the burst errors. Differentially demodulated phase shift keying presents a special case, in that there is a small irreducible background of errors at any signal-to-noise ratio, owing to abrupt phase changes caused by the propagation medium.

With non-Gaussian noise, particularly of the impulsive character associated with man-made noise, there is a further tendency to bursts of errors. The bursts owing to impulsive noise are, however, not closely related to the signal fading pattern because of the comparative insensitivity of impulse induced errors to changes in signal level.

9.6.2   Types of Coding

Time diversity for digital data is a simple technique which involves repeating the message, or parts of it, at different points in time. Delays of the order of 100 ms should be suitable for spanning fast, deep fades where an adaptive serial modem is most likely to lose track of the HF channel. This kind of diversity can be incorporated into the message to reduce burst error problems.

An alternative and more powerful method of improving channel performance is by the use of coding techniques. Coding schemes have been evolved which form an alternative to diversity working when using system or data redundancy to improve performance. Instead of sending the message itself twice, for example, a code word is sent. This corresponds to a specific segment of data which may enable the receive system to determine, not only when an error has occurred, but also where that error has occurred, thus allowing correction of the error.

Many error correction, detection and control schemes have been devised, each claiming different abilities to contain or control error occurrences. There are three different ways in which redundant information may be used to improve the accuracy or reliability of the received information:

1 *Error Detection* Typically a code with about 10% redundancy can be used to detect all cases of a single element error in a block and some cases of multiple errors. However, some cases of multiple errors will be undetectable.

2 *Error Correction* By using a larger amount of redundancy, typically 40%, it is possible to correct single errors in a block and sometimes a limited number of double errors.

3 *Automatic Repeat Request (ARQ)* This is an error correction system in which the data is coded with redundant information so as to detect certain classes of error pattern. Detection of an error will initiate automatically a request for repetition of some or all of the data.

Error control coding can be used for forward error correction (FEC) or for automatic repeat request (ARQ) operation. FEC requires no return path, and employs error correction coding to improve performance. In ARQ operation error detection coding is used, and the detection of an error initiates a repeat request, which is sent over the return path.

In some circumstances it is appropriate to use error detection rather than error correction even when there is no return path over which a repeat can be requested. This applies, for example, when the messages passed are teleprinter messages in ordinary text. The detection of an error in a teleprinter character can be caused to print a special error symbol. The message can then often be interpreted reliably even in the presence of a few errors, because of the redundancy inherent in the language. Another instance is in telemetry, where the complete loss of an occasional reading in a continuous flow of slowly-varying readings may be unimportant, but an undetected error in a reading would be serious.

ARQ operation is potentially the most efficient way of conveying data with very low error probability. This is because error detection is used, which requires less redundancy to control a given channel error probability than does error correction; in addition the system is adaptive, because repeats only occur when needed. It is possible to introduce a further level of adaptation by using the repeat rate as a measure of the difficulty of transmission. If the repeat rate rises above a specified level, steps can be taken to reduce it, by, for example, reducing the transmission rate. In the absence of adaptation of this sort, ARQ systems are liable to 'block up' completely in both directions if the channel error probability rises above some critical level. Moreover, for some HF circuits, when one direction is operating successfully and the other direction is showing high error rate, the 'good' direction could be interrupted severely by the continual repeats.

The implementation of an error control coding scheme usually takes the form of a 'block' code or a 'convolutional' code; each is now discussed briefly.

9.6.3   Block Coding, with Hard Decision

Consider systems in which each digit has one of two states, 0 or 1, and in which the receiver makes a firm decision as to which has been received. In *block coding* the message bits are, before transmission, divided into blocks of k digits, thus there are 2^k possible distinct message blocks. The coding is applied to each block individually, to produce a block of n digits; this is done by adding $(n - k)$ additional digits (so-called *parity digits*) which are chosen in accordance with the particular combination of k message bits in the block. It is the logical process by which the parity bits are produced which constitutes the code. A code is said to be *systematic* if every code word contains the corresponding k message bits, followed by the $(n - k)$ parity bits.

There are 2^n possible distinct blocks of n digits. However, the only ones ever transmitted are the members of the sub-set of 2^k blocks. Each member of this sub-set corresponds to one of the 2^k message blocks. If a block of n digits is received which does not correspond to one of the 2^k encoded blocks, then it is known that errors have been introduced during transmission. (It is assumed that the encoded block transmitted was the block which differs least from that received, in terms of the number of digits difference.) The corresponding k message bits are then passed to the decoder output. This process is described as *error correction*. Note that this method of decoding is only optimal if all the possible message blocks are equally likely to be transmitted. This will be the case if, for example, the data is encrypted.

If the received block differs from two or more valid encoded blocks by the same number of digits, then the encoded block transmitted cannot be identified, but it is known that the received block is invalid. This is *error detection*. In the usual nomenclature, a block code is defined as an (n,k,t) code, where n and k have the meanings given above, and t is the number of errors in the encoded block which are correctable. The *minimum distance* of a code is the number of digits difference between the two code words whose difference is the minimum. A code with minimum distance d can correct up to $\frac{1}{2}(d - 1)$ incorrect digits, or detect up to $(d - 1)$ incorrect digits. Thus,

$$t = \tfrac{1}{2}(d - 1) \tag{9.12}$$

It is possible to combine error correction and error detection. Alternatively, the decoding process can be constrained so that the number of errors corrected is less than that permitted by the minimum distance; the distance between code words unused for error correction is then available for error detection.

The efficiency of a code is expressed in terms of the *code rate*, which is usually given the symbol R, where $R = k/n$. The art of code design lies in providing a given error correcting or detecting power with the maximum possible code rate, without requiring encoding or decoding processes which are excessively difficult to implement.

9.6.4   Convolutional Coding, with Hard Decision

Convolutional coding operates on the information sequence without breaking it up into physical blocks; it is performed in the following way. The message digits are passed into a shift register in blocks of k digits. The shift register is long enough to hold h such blocks at a time, thus the total length of the register is hk digits. By a logical combination of the hk digits in the shift register, n digits are formed and fed to the encoder output. The next k message digits are then fed into the shift register, the oldest k digits are discarded, and another block of n output digits is generated from the total contents of the register, and so on. It should be noted that the number k is usually quite small, and often unity.

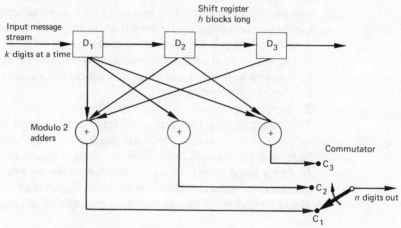

Fig. 9.19 *A simple convolutional encoder*

Consider a simple example in which $h = n = 3$ and $k = 1$ (see Figure 9.19). During each message bit interval, the commutator samples the modulo 2 adder outputs C_1, C_2 and C_3. Thus a single message bit yields three output bits. The next message bit in the input sequence enters D_1, while the contents of D_1 are shifted to D_2 and D_2 to D_3; the commutator again samples the three adder outputs. In this way each of the k input digits has an influence on $n \times h$ output digits.

The above procedure generates n encoded digits for every k message digits; n is always greater than k, and the code rate $R = k/n$. The shift register length hk is called the *input constraint length*, and hn is called the *output constraint length*. On receipt of the message in hard decision decoding, the receiver demodulator makes a firm decision regarding each received digit.

A more detailed explanation of convolutional codes can be found elsewhere[19]. In this chapter, further discussion of error control coding is limited to the simpler block codes.

9.6.5 Detection and Correction of Error Bursts

The limit of the error correcting or detecting power of a block code is some specified number of errors per block. If the errors occur in bursts, the code will fail at an average error probability which the code could control if the errors occurred randomly. (Convolutional codes are subject to a similar limitation.) There are three different approaches available for overcoming this difficulty:

1   *Long Block Approach*

The block length (for a block code), or the constraint length (for a convolutional code), is made very long. Ideally the length should be such that the number of errors in the longest burst expected will represent only a small proportion of the errors correctable within the block. This is often impracticable, and it is then necessary to be content with a block or output constraint length capable of accommodating the maximum number of errors expected in a single burst. With such an approach it is possible to use codes designed specifically to correct burst errors. These codes take advantage of the fact that to correct a burst of t errors in a block of n digits requires less information (in the 'informaton theory' sense) than is required to correct t bits randomly distributed, thus a higher rate code can be used.

The adoption of such a code necessarily reduces the ability to correct random errors, so if both burst and random error correction are required the code must be chosen with that in view.

2   *Symbol Approach*

The message may be segmented into sections, and the sections treated as symbols. The message is then encoded in terms of symbols. It follows that a burst of digit errors falling within one section appears as a single symbol error. In fact a burst might fall across the boundary between two symbols, so it is necessary to use a code capable of correcting at least two symbols. The Reed-Solomon codes[20] are the principal example of codes of this type currently in use.

3   *Interleaving Approach*

The burst errors may be spread in time by *interleaving*. At the transmitting end the encoded digit stream is loaded into a rectangular matrix row by row. The digits are then read out for transmission column by column (see Figure 9.20). At the receiving end the digits emerging from the demodulator are loaded into a matrix column by column and read row by row into the decoder. By this process, errors which occur in bursts are distributed at intervals along the digit stream entering the decoder.

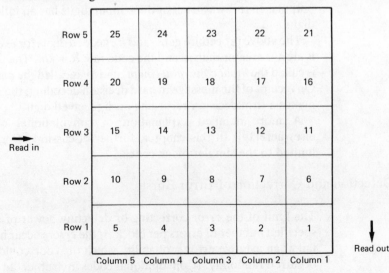

Fig. 9.20   *A 5 × 5 interleaving matrix*

All three of the above methods entail a delay in the delivery of the message. This is necessarily the case, because an error burst arises when the channel is, for a period, incapable of conveying the required information. All the methods of coding described overcome this by repeating the information at another time, or by spreading the information over an interval long compared with the duration of the channel disturbance. This is, of course, time diversity.

9.6.6 Soft Decision Decoding

It has so far been assumed that the demodulator makes a firm decision regarding each received digit. In the binary case the decision is either 0 or 1. The simplest departure from this is *erasure decoding*. If a demodulated signal is closer to the decision threshold than some specified limit the digit is called an erasure, and no value is assigned to it.

This approach can be used in conjunction with an error detecting code. Provided that the number of erasures does not exceed the number of errors which the code can detect, the correct digit sequence can be found by assigning, in turn, all the possible combinations of values to the erasures, until a combination is found which gives zero detected errors. Since a code with minimum distance d can detect $(d - 1)$ errors but can correct only $\frac{1}{2}(d - 1)$ errors this is obviously an advantage.

More complex systems can be used, which account for the amplitude of the demodulated signal in relation to the decision threshold, and assign a probability to each decision. These systems use what is generally known as *soft decision decoding*[21], of which erasure decoding is the simplest example. The value of soft decision decoding and erasure decoding differs according to the type of noise environment within which the communications system must operate.

9.7 Performance of Error Control Coding

9.7.1 Effect of Error Probability

Any coding scheme reduces the effective message throughput and this must be traded off against improved link performance; methods to overcome burst errors necessarily entail a delay in the delivery of the message. It is characteristic of coding over HF channels that it is much easier to produce a very low output error probability from a mediocre input value than to produce a mediocre output from a poor input value. With a sensible choice of code, valuable improvements may be made in link quality. The overall improvement achievable by a given code is greatest when diversity is not present in the system, because diversity schemes themselves reduce error rate.

If error control coding is applied to a data channel, while keeping constant the transmission speed in message bits per second, there is a reduction in energy per signalling element transmitted, because of the redundancy introduced by the coding process. This causes an increase in the error probability of the received signal elements, which tends to offset the advantage given by the code. Because of this effect it is necessary, when choosing a method of error control, to take into consideration the error probability versus E_b/N_0 curve of the system, as discussed in section 9.3.

9.7.2   Coding Gain

Consider as an example, FSK modulation with non-coherent reception, and the effect upon such a signal of applying four different error control codes:

(3,1,1) majority vote code, code rate 0.33
(15,11,1) Hamming code, code rate 0.73
(23,12,3) Golay code, code rate 0.52
(63,45,3) BCH code, code rate 0.71

The benefit given by coding is usually expressed as a *coding gain*. This is the reduction in required signal power for a given error probability when coding is in use, compared to the signal power required for the same error probability without coding. The coding gains given by the four representative coding schemes are listed in Table 9.3, for the various situations considered in Figures 9.21 to 9.23. In these examples it is assumed that interleaving is used, so that the error distribution at the decoder input approximates to random. Methods of calculating error probabilities for block codes in Rayleigh fading without interleaving are described elsewhere[22].

	CODING GAIN IN DECIBELS							
	(3,1,1) $R=0.33$		(15,11,1) $R=0.73$		(23,12,3) $R=0.52$		(63,45,3) $R=0.71$	
ERROR PROBABILITY	10^{-2}	10^{-4}	10^{-2}	10^{-4}	10^{-2}	10^{-4}	10^{-2}	10^{-4}
Non-fading	−2.1	−2.0	−0.3	0.7	−0.1	1.0	−0.1	1.5
Rayleigh fading, no diversity	3.5	12.9	2.7	12.0	5.0	19.0	3.0	16.9
Rayleigh fading, double diversity	0.3	4.5	0.9	5.6	1.2	8.1	1.9	8.8
Rayleigh fading, quadruple diversity	−1.1	0.8	0.3	2.7	0.7	4.4	0.4	4.4

Table 9.3   Theoretical coding gain for various coding schemes

9.7.3   Analysis of Examples

In Figures 9.21 to 9.23 curves are given based on a majority vote; it is assumed that a repetitive message is sent and some simple decision-making circuit is present.

A study of Table 9.3, and Figures 9.21 to 9.23, reveals the following information by considering error probabilities of between 10^{-2} and 10^{-4}:

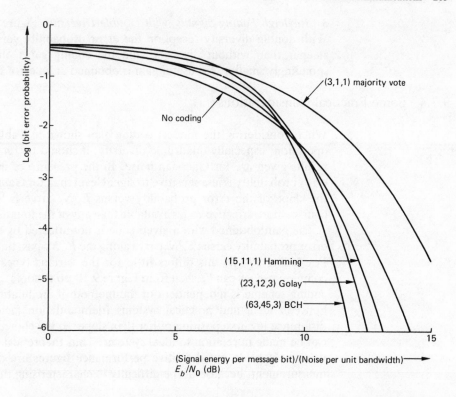

Fig. 9.21 *Effects of coding schemes on non-coherent FSK with non-fading signals*

1 *Non-fading Signals* (Figure 9.21)

For non-fading signals all the codes considered are quite ineffective at an error probability of 10^{-2}, and in fact degrade the performance relative to that obtained without coding. At an error probability of 10^{-4} three of the codes provide a small benefit, but not enough to be worthwhile. This is because of the large slope of the error probability versus E_b/N_0 curve before coding, for non-fading signals.

These results are for Gaussian noise; in the presence of impulsive noise the value of error correction with non-fading signals becomes evident, especially if low output error probabilities are required. Since non-fading signals are not relevant to HF sky wave communications they are not considered further.

2 *Rayleigh Fading Signals* (Figure 9.22)

All the codes give an improvement at both error probabilities considered, but the coding gain is much greater at 10^{-4} than at 10^{-2}. This illustrates the fact that it is much easier to produce a very low output error probability from a mediocre input value, by coding, than to produce a mediocre output from a poor input.

Because of the small slope of the error probability versus E_b/N_0 curve before coding, the increase in the signalling element error probability due to the redundancy introduced by coding is not severe. The (23,12,3) code (rate 0.52), performs better than the (63,45,3) code (rate 0.71), or the (15,11,1) code (rate 0.73). The (3,1,1) code has an even lower rate of 0.33, but this is offset by the low effectiveness characteristic of very short codes.

3   *Rayleigh Fading Signals, with Double Diversity* (Figure 9.23)
With double diversity reception the error probability versus E_b/N_0 curve is steeper than without diversity, and the coding gains obtained are correspondingly smaller. No useful gain is obtained at an error probability of 10^{-2}.

9.7.4   Some Practical Considerations

When considering the modest coding gain shown in Table 9.3 for diversity operation, especially quadruple diversity, it should be remembered that the results given are for Gaussian noise. In the presence of impulsive noise the error probability is less sensitive to signal level than for Gaussian noise, so that the slope of the error probability versus E_b/N_0 curve is less. There are, of course, more effective codes available than any of the four used for illustration.

The gain obtained with a given code is not affected by the position of the error probability versus E_b/N_0 curve along the E_b/N_0 axis, but only by the slope. For fading signals this differs little for the various types of modulation in common use, as can be seen from Figure 9.10, so to some extent the choice of coding scheme is independent of the method of modulation. The difference between ideal and practical systems (demodulation factor) is primarily a difference in curve position rather than slope, so the choice of coding method can be made in relation to ideal systems. Thus theoretical calculations are of value as a guide but definitive performance figures are only obtainable by measurement, because of the difficulty in characterising the noise.

9.7.5   Time Delays

Using error control coding inevitably causes some delay, because information which would have been conveyed by one transmitted symbol is spread over more than one symbol; the delay increases with the length of the code block used (or with the output constraint length for a convolutional code). In the presence of fading, intermittent interference or noise bursts it may be necessary to spread the information deliberately over some minimum time to obtain time diversity.

There is also a delay caused by the encoding and decoding processes; in practice decoding takes much longer than encoding. Because of the modest data rates which can be used over HF channels, and the very high speeds obtainable with state-of-the-art digital circuits, encoding and decoding time will probably not be found to be too important. It must not be ignored completely, however, because some decoders have a decoding time which increases very rapidly with code length and correcting capability.   For a block code the delay incurred by using the code (excluding the encoding and decoding time) is the time taken to send a block. This is because no use can be made of the first digit until the last one in the block has also been received. (Serial input and output of data is assumed.) Thus

$$\text{Delay} = (n/B) \text{ seconds}$$

where n = number of digits per block
$\quad\quad B$ = number of digits transmitted in one second.

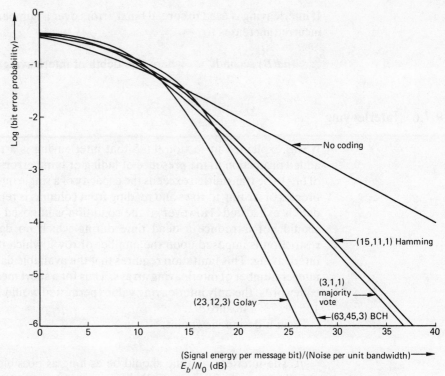

Fig. 9.22   *Effects of coding schemes on non-coherent FSK with fading signals*

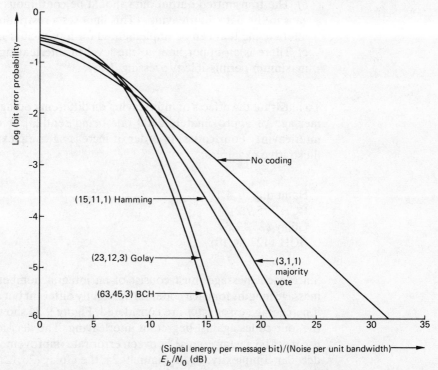

Fig. 9.23   *Effects of coding schemes on non-coherent FSK with fading signals with double diversity*

If interleaving is used to spread burst errors over a number of blocks, the delay incurred increases to

$$(nd/B) \text{ seconds} \qquad \text{where } d = \text{depth of interleaving}$$

9.7.6   Interleaving

It was explained in section 9.6.5 that interleaving is a method used to gain added protection in the presence of fading or burst errors. If the total number of bits to be transmitted exceeds the capacity of a single interleaving matrix, the process of writing to rows and reading from columns is repeated until the input data is exhausted. However, if the condition is imposed that the interleaving should not introduce a dead time during which no data is transmitted, a restriction is imposed upon the number of rows, which defines the degree of interleaving. This limitation requires that the available data fits exactly into an integer number of interleaving arrays. Thus for a short message comprising 292 codewords, the only interleaving values permitted would be factors of 292, i.e. 1, 2, 4, 73, 146 and 292.

From a design viewpoint:

a) The interleaving time should be as long as possible (but limited by the fading period of the channel).
b) The transmitted output data should be continuous so there must be no gaps in the data interleaving. (This imposes a restriction upon the levels of interleaving that can be implemented for a particular message duration.)
c) There is an upper limit to the level of interleaving, resulting from the maximum permissible processing delay.

To illustrate the effects of interleaving on different coding schemes, consider a message of approximately 2000 bits being sent[23] at different degrees of interleaving. Four codes, in order of increasing complexity, are examined as illustrative examples:

Hamming (7,4,1)
BCH (15,7,2)
Golay (23,12,3)
BCH (127,64,10)

Since each message must consist of an integral number of codewords, the message lengths for each code will be slightly different but a comparison is valid if percentage error rates are considered. Figure 9.24 shows the comparison of the four codes against degree of interleaving. The decision on which code to use must be a compromise between error rate improvement obtained, and the decoder complexity. From Figure 9.24, the Golay (23,12,3) code is preferred, since it gives almost as many error free messages as the BCH (127,64,10) code (at their maximum interleaving levels for continuous transmission).

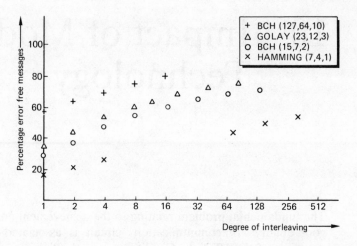

Fig. 9.24 *Effect of interleaving on error rates for different coding schemes*

9.7.7   Erasure Decoding and Soft Decision Decoding

In a Gaussian noise environment soft decision decoding gives a worthwhile improvement in terms of the reduction in required signal power for a given error rate. The symbol reliability information supplied by the demodulator must however define the relative probability of correctness of symbols to within fairly fine limits. If only one bit of reliability information is used (i.e. erasure decoding), the improvement over hard decision is so small as not to justify the additional complexity incurred.

In the presence of impulsive noise, interference, and fading, which is the relevant context for HF radio, erasure decoding can give a substantial improvement in performance; moreover the use of full soft decision is unlikely to give sufficient further improvement over erasure decoding to justify the increased complexity.

Erasure decoding is particularly effective if the presence of strong interference, noise bursts, or deep fades can be detected. The symbols known to be affected can then be declared as erasures. The decoder (if of a suitable type) uses up less of the available power of the code in correcting a symbol which has been declared as doubtful, than it uses to correct a symbol whose error it is left to find out for itself.

10 Impact of Modern Technology

10.1 Overview

The fundamental problem relating to the achievement and maintenance of a good-quality HF communications circuit is associated with the inherent dynamics of the HF link. Conditions over the link may change rapidly; the communications systems must be able to adapt to the prevailing conditions.

The dynamic nature of the HF medium necessitates the use of a fast processing capability to optimise the system performance. Digital radio frequency and control system technology together with a cheap and reliable fast processing capability combine to address the HF problem in an achievable way. Digital control systems can provide the nervous system for modern HF systems design. They can orchestrate automatic adaptation, timing, frequency hopping, radiated power level control, information flow control and other control functions required in a modern HF adaptive system. Techniques for real time frequency management are now being implemented in current systems.

The impact of very large scale integration (VLSI) will be evident in all types of future communications sytems. VLSI technology provides the ability to reduce size, weight and power consumption of existing equipment and to increase the overall system reliability. It also enables more complex systems to be provided by introducing techniques and applications which have not previously been possible because of the lack of necessary technology.

Highly reliable HF systems are dependent upon efficient signal processing. Sophisticated digital signal processing has made new modulation modes practicable, whilst at the same time reducing the equipment cost. Such techniques operate on the modulating signal or on the RF carrier in a manner which increases the quality of intelligence transferred from the sender to the receiver. VLSI techniques enable sophisticated modulation and coding to be implemented; high levels of data interleaving can be used to overcome the effects of burst errors. Digital modems are adaptable to different modes by software techniques, avoiding hardware retrofits or conversions. The modulation and demodulation processes can now be performed digitally. Wideband modulation schemes and frequency hopping techniques are becoming more common; digitally controlled synthesisers are capable of changing frequency many times a second.

Voice processing and speech recognition techniques enable limited vocabulary speech to be transmitted over very narrow bandwidths. Applications that rely upon fast processing of voice signals demand the use of filters that can provide high stability and performance. This can be achieved through the use of VLSI technology in the implementation of digital filters.

Progress in VLSI techniques is such that integrated receiver front ends, incorporating frequency changers and IF amplifiers, are already feasible. Future developments may lead to direct analog-to-digital conversion of RF signals with high rate conversion. Problems of dynamic range, intermodulation and conversion speeds remain to be overcome.

The use of electronically steered receiving antenna arrays is also made feasible by the use of VLSI techniques which are able to provide the processing power necessary to perform efficient wavefront analysis. A null in the radiation pattern can be steered electronically to minimise the effect of an interfering signal on channel performance.

Evolving equipment designed to meet continuing needs has yielded significant improvement of receivers, transceivers and input/output devices. Equipment that relieves onerous tasks such as radio tuning and radio monitoring is now popular. The technical challenge has been to obtain electronic control without losing the performance available from the best manual equipment.

While digital technology will assume some of the hardware burden, high power and high frequency analog circuitry will still be required. The key thrusts are in tuning speed, spectral purity and system bandwidth.

Although future systems will undoubtedly be considerably more complex that those at present on the market, VLSI technology will allow new equipments to occupy the same volume and have somewhat similar weight and power consumption requirements as current communications transceivers. Fault detection and diagnosis can be improved using processor-based monitoring, control systems and built-in test equipment (BITE). These enable on-line system tests to be automatic and repeated on a frequent basis to indicate potential problems before they can become serious.

Even greater capabilities will result from the signal processing power of very high speed integrated circuitry (VHSIC) technology. The associated speed and processing power will replace additional circuit boards of analog components in the processing of still more sophisticated waveforms not possible today. The trend will continue with digital embedding of present day functions such as artificial intelligence-based control and cryptographic facilities.

The scope for technology in improving HF communications systems appears to be considerable. The aim of this chapter is to discuss some of the principal trends for future HF communications systems. These advanced techniques, as will already be evident from this brief overview, cover a wide range of technical innovation.

For the convenience of description in this chapter key advanced techniques are categorised in terms of:

Hardware Evolution
Signal Processing
Interference Reduction
Adaptive Techniques
Speech Processing

10.2   Hardware Evolution

10.2.1   Receivers

The world's first HF communications receiver to use digital signal processing for all IF and baseband functions was introduced in 1984. Crystal filters, trimmer capacitors and IF transformers have been replaced by read-only memories and microprocessors for all IF and baseband functions.

Previous-generation receivers used narrowband mechanically-tuned radio frequency amplifiers and local oscillators. New receivers tune local oscillator synthesisers under digital control. They do not use narrowband RF preselection because this introduces considerable non-linearity which causes an increased noise floor composed of high-order intermodulation products. New receivers are operationally very convenient. Digital frequency selection enhanced by the ability to store frequencies provides an important reduction in operator workload. Many receivers use push button or rate-of-scan tuning to aid manual frequency selection. Functions are controllable by a microprocessor at a remote location and check-out is under software control (built-in self-test, BIST).

Although new frequency-agile receivers exhibit good dynamic range, the noise figure suffers. An additional problem specific to receivers co-located with transmitters is that of receiver front-end protection. The problem of simultaneous transmission and reception is particularly acute on a small platform such as an aircraft. The difficulty arises because of the close coupling between transmit and receive paths, which is inevitable given the small size of aircraft relative to the wavelengths involved. Isolation of transmit and receive antennas is physically impossible. Thus significant voltages appear on receiver inputs during transmitter operations; these cause problems of intermodulation, blocking and reciprocal mixing[1]. However, although substantial signals are present (possibly tens of volts), these voltages can be handled by solid state tuning components. The use of front-end solid state filter units, possibly operating under digital control, could permit simultaneous operation, providing that minimal frequency separation is maintained.

10.2.2   Transceivers

New transceivers are following the lead of digitally controlled receivers. Computer terminal control via a standard telephone interface is becoming more common; it is available also in radio transceivers for amateur use. Transceivers continue to be designed primarily with SSB modulation. This provides a voice bandwidth channel from which the interface to digital data peripherals can be specified. The SSB radio practice has been to emphasise sharp amplitude response filters which pass the 300 Hz to 3 kHz frequencies and heavily attenuate frequencies outside this region. The channel filters cause particularly large phase shifts near the band edges. This deviation from a linear phase/ frequency causes 'ringing' leading to intersymbol interference when receiving serial data and to degraded bit timing on parallel tone data. More recent transceiver specifications now define band-pass characteristics with data transmission requirements in mind.

10.2.3   Power Amplifiers

Many power amplifiers in current use employ vacuum tubes. Output stages are tuned to minimise harmonics, intermodulation products and other spurious and wideband noise outputs. Tuning is typically accomplished through use of servo systems which are electromechanical in nature and are therefore relatively slow (taking typically 10–30 seconds to tune across the HF band). They are less reliable than semiconductor circuitry.

With the development of RF power transistors, a new linear RF power amplifier architecture has emerged whereby the outputs of a number of relatively low power modules are combined using 90° hybrid couplers to produce a high power unit, typically of several hundred watts. The system is wideband throughout, permitting essentially instantaneous frequency selection.

Modern power amplifiers do not suffer damage from output mis-match. Circuitry can now be provided to sense the mis-match and turn off the input signal before damage occurs.

10.2.4   Synthesisers

The two important parameters for programmable frequency synthesisers are tuning speed (settling time) and spectral purity (particularly off-tune residual phase noise). *Tuning speed* sets the limit on channel scanning (or hopping rates in frequency hopping systems) while the *spectral purity* determines the selectivity of receivers operating in a high interference environment. Off-frequency interference causes a *reciprocal mixing* problem even if transmitter signals are pure tones. (This occurs as a result of off-tune synthesiser noise signals mixing with off-tune receiver input signals to produce noise at the receiver IF frequency, see Figure 10.1.)

Two basic kinds of synthesiser have been developed. The indirect synthesiser uses phase-lock loops and thus has a relatively long settling time ($\approx$ 1 ms or more). The direct synthesiser uses mixing/filtering processes which settle quickly ($\approx$ 10 μs). The relatively poor noise performance of early direct synthesisers has now been largely overcome. However, the direct synthesiser is still significantly more costly than the indirect type. Receiver reciprocal mixing in a high noise environment is still a problem.

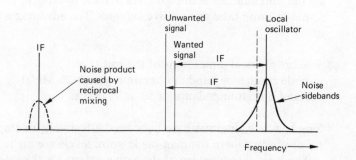

Fig. 10.1 *Reciprocal mixing*

10.2.5   Transmitting Antennas

Efficient operation at HF generally requires the use of a relatively large transmitting antenna. However, a fully adaptive HF system can usually operate on low power and does not require efficient antenna structures. Broadband antennas, providing a good impedance match over the band, are either very large or very inefficient as a result of the use of resistor loading to improve the impedance match. Transportable systems can avoid the need for the broadband antennas by using antenna couplers but these are suitable only for tuning single-ended antennas such as whips. Transportable and mobile antennas have advanced little in the last decade beyond the automatic tuning of narrowband antennas. Antenna coupler tuning speeds of a fraction of a second are obtainable in units which switch vacuum relays to pre-determined positions. However, rotary mechanisms taking seconds to tune are common. In both cases the poor reliability of mechanical components can cause radio maintenance problems.

Antenna tuning unit (ATU) tuning speeds can be improved by pre-programming of ATU settings. This technique also provides silent tuning. However, the need to use mechanical components to cater for high transmitter powers limits tuning speeds to a few tens of milliseconds. Solid state ATUs, using PIN diodes, possibly combined with integral power amplifiers, are a possible development. They have already been demonstrated to be capable of handling a few hundred watts of input power.

10.2.6   Receiving Antennas

For most systems, reception requires greater agility than does transmission. Receiving antennas can be made broadband through impedance matching and active elements. There are now many commercial and military versions of the short electrical antenna coupled by transformer and/or amplifier to a transmission line.

Because relatively inefficient antennas can be used for reception without detriment to overall system performance, advantage can be taken of this fact to use electrically small receiving antennas either to facilitate the rapid deployment of tactical HF terminals or where there are limitations on available ground area.

An active antenna consists of a rod or dipole and an RF amplifying device. If the noise level generated at the amplifier terminal is less than the noise pickup by the antenna, an active antenna system is capable of supplying the same signal-to-noise ratio as a passive antenna. The advantages of an active antenna are:

Short physical dimensions of the rod
Wide frequency band of operation (i.e. 2–30 MHz)
Fixed output impedance of 50 or 75 ohms.

Simple active antennas have been used effectively for shipborne applications, for example, where the man-made noise levels are high.

The main disadvantage of the active antenna is the intermodulation distortion products which are often generated by strong interfering signals. These distortion products can degrade the performance of an active antenna in an environment with high RF noise levels. The linearity of the active circuitry must be as high as possible, to minimise the level of intermodulation products.

10.3   Signal Processing

10.3.1   Lincompex

Signal processing techniques for HF have been in use for many years. Lincompex[2] (LINked COMPressor and EXpander) is a signal processing technique adopted as a standard by the CCIR[3] after its original development by the British Post Office. The basic principle is that variations in the level of the input speech, particularly those caused by inflection, expression and syllabic variation, are compressed to a constant amplitude for transmission. They are expanded again upon reception to the original levels under the operation of a separately transmitted control. This system is now integrated within HF radio links in many telephone networks, providing standards of quality and stability which approach those of cable and satellite systems. Recent developments in miniaturisation now make this HF performance available to mobile units and open a whole new field of applications in long-distance reliable communications. Modern Lincompex systems give the opportunity to integrate the HF radio link completely within existing communications networks. Mobile HF stations can now be an integral part of the extended lines of communication without relying on vulnerable relay points. The Lincompex technique has been available for some twenty years, and much more sophisticated techniques based upon digital signals are now becoming possible.

10.3.2   Digital Receive Modules

Progress in VLSI techniques is such that direct analog-to-digital conversion (ADC) of RF signals with high rate will become possible, but considerable technical problems of dynamic range, intermodulation, conversion and processing speeds remain to be overcome. An alternative approach is the use of direct conversion to baseband (zero IF) with simple low pass filters (to avoid aliasing[4]) followed by analog-to-digital conversion. However, this approach has so far been unable to achieve the performance of conventional HF receivers, and long ADC word lengths are required to cater for the high dynamic range of signals entering the receiver front-end. Hence it is probable that, for some time into the future, VLSI receiver devices will combine conventional frequency conversion circuitry with on-chip frequency synthesisers and digitisation of band-limited IF signals.

Figure 10.2 illustrates one possible configuration for a digital active receive element. A conventional active antenna feeds a number (four in the example of Figure 10.2) of independent VLSI frequency convertors, which produce quadrature outputs at a common IF. Each convertor has an independent synthesiser, locked to a common frequency reference, and can accept control signals via a high-speed data bus to select operating frequency, IF bandwidth, AGC level, digital sampling rate and other parameters as required.

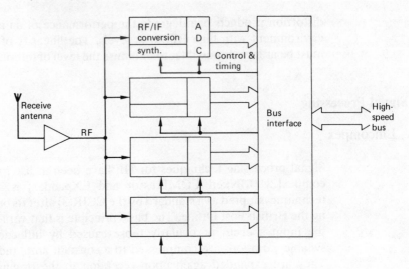

Fig. 10.2   *Active digital receive module*

10.3.3   Signal Detection

Simple signal detection schemes based upon input signal level (squelch) are
rarely applicable at HF. This is because of the great variability of anticipated
received signal strength caused by operating frequency, and diurnal and seasonal ionospheric variations. An automatic method for signal detection must
therefore provide a means of detecting changes in a range of characteristics of
the received signal. The most robust technique appears to be straightforward
spectral analysis using fast Fourier transforms (FFT). HF channels are rarely
completely quiet in the absence of wanted signals; the spectral analysis routine
must compare the background spectrum of the residual interference with that
caused by the wanted signai.

Frame lengths for the FFT must be sufficiently long to provide adequate
resolution of narrowband wanted and unwanted signals, but should not be so
long that undue delay is incurred in detecting the presence of a wanted signal.
Frame lengths between 40 ms and 100 ms are likely to be suitable for data
signals but 500 ms may be required to cater for gaps in speech. Many efficient
algorithms are currently available for FFT implementation. A fairly high false
alarm rate in the initial detection of an incoming call may well be acceptable,
since subsequent processing can be used to confirm the presence of a wanted
signal. The FFT analysis has therefore only to act as a coarse filter and no
elaborate processing is required.

10.3.4   Signal Identification

The task at this stage of the processing is to identify unambiguously the type of
signal, and to confirm that the message is addressed to the particular recipient.
In the case of narrowband modulation such as FSK or MFSK, the information

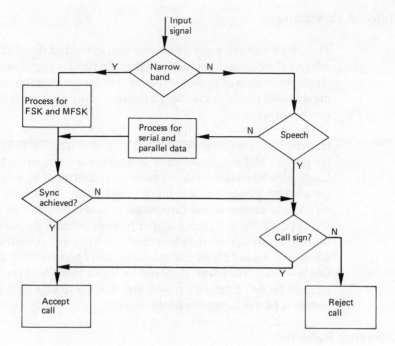

Fig. 10.3 *Decision tree for decoding incoming message*

from the spectral processing may be sufficient to identify the modulation scheme. For broadband signals, such as 2400 bit/s PSK (serial or parallel tones) or speech, there seems little option but to process the signal, in parallel, for all possible modulation schemes, and to depend upon successful decoding of the message information. A simple decision tree is shown in Figure 10.3 as an example.

Digital techniques for demodulation of narrowband low data rate signals are straightforward and do not impose a high computational workload. For MFSK demodulation either digital filtering, to generate a filter bank, or FFT processing may be used. For conventional two-tone FSK systems, digital filtering is probably more economical. In both cases, reduced sampling rates can be used if the demodulation process is preceded by a narrowband digital filter.

Broadband signals (speech or data) present the highest processing loads. Processing loads of up to 15 million instructions per second may be required for demodulation of 2400 bit/s data (serial or parallel tones) if adaptive equalisation is required. Use of broadband, computer-controlled receiver elements permits the introduction of frequency management schemes dependent upon fast tuning, to enable a receiving station to monitor a wide spread of transmissions from a number of stations. The signal analysis and demodulation techniques may be applied to signal identification and call processing, thus eliminating the need for continuous monitoring of noisy HF channels.

10.3.5   Transmit Processing

The scope for major enhancements in transmitted signal facilities by applying advanced techniques is much less than for receive signal processing. However, direct generation of signals at transmitter drive unit levels using digital processing is feasible and requires only a fraction of the processing load associated with receive processing.

The major restriction in implementing more flexible procedures for radio transmission from large HF stations is the requirement to handle high transmitter powers and the consequent impact upon antenna and antenna couplers. Large ground stations will require digital control of antenna selection and the use of high-power multiple-input linear power amplifiers, attached to omni-directional antennas and directional transmit arrays. The approach provides the opportunity to transmit higher powers, when only one or two services are sharing a transmitter, than is possible with current ground station installations which use a mix of high and medium power transmitters, each dedicated to a single service. However, it should be noted that linear transmitters are inefficient in terms of primary power and that there may be a problem with total installed power capability of the system.

10.4   Interference Reduction

10.4.1   Interference Nulling at the Antenna

By summing the outputs of a number of individual antenna elements, with appropriate amplitude and phase weighting, radiation patterns can be produced which favour the wanted signal by rejecting interference. This is achievable by directing nulls or minima in the radiation pattern in the direction of the interfering sources. An array of n elements can, in principle, provide $(n - 1)$ independently controllable nulls. In practice arrays often possess symmetry; this can result in the number of degrees of freedom being effectively fewer than would be expected from the actual number of elements.

Natural and man-made interference arriving from a different azimuthal direction to that of the wanted signal can be reduced significantly in amplitude by suitable phasing of the array to produce nulls in the radiation pattern of the receiving antenna. A similar principle can be applied in the elevation plane. Multi-mode propagation may produce two signals of comparable amplitude arriving at different elevation angles; this results in deep fades. A null in the antenna radiation pattern directed towards one of these components will reduce the signal fluctuations considerably. It is unlikely that such a technique will be effective in reducing within-mode fading, owing to the small angular separation between the components of each mode. Nevertheless, adaptive arrays should be effective in attenuating unwanted transmissions and weaker ionospheric modes which arrive from directions well separated in the angular domain from the main signal.

Interference nulling can be regarded as a limited version of signal-to-noise ratio optimisation provided by adaptive arrays, discussed in section 10.5.3. An array is given a basic radiation pattern appropriate to the reception of the required signal, and an algorithm then operates to modify the pattern to minimise the levels of discrete interference sources. In its simplest form the method uses two omnidirectional antennas, for example vertical whips. The cancellation algorithm operates to minimise the interfering signal level. It is necessary to ensure that the unwanted signal is cancelled rather than the wanted signal and this can be achieved by a number of different methods:

a) The adjustment of weights can be done by an operator, who distinguishes the signals by ear.

b) An algorithm which operates to minimise the receiver input will cancel the larger input signal, so if the unwanted signal is the larger it will be cancelled.

c) If the spectrum of the unwanted signal extends beyond the necessary bandwidth of the wanted signal, then the input to the adaptation algorithm can exclude the wanted signal. This is particularly useful in jammer rejection, because jamming signals are often fairly wideband.

d) The adaptation process can be made to operate only when the wanted signal is absent, the array weights being locked when the wanted signal is present.

e) If the wanted signal does not occupy all its bandwidth all the time, for example with a two-tone data signal, the adaptation can be based upon the part of the bandwidth not occupied by the wanted signal at any instant.

The two-element array described above provides a basically omnidirectional pattern in azimuth, to which an interference-rejecting null is added (a second mirror image null is necessarily also produced). The introduction of the null causes some modification to the gain in the direction of the wanted signal, and if the angular spacing between wanted and unwanted signals is too small the wanted signal will be attenuated unacceptably.

The same approach can be used with more complex arrays. A multi-element array can be steered open loop, to give a gain in the known direction of the wanted signal. The signal from a single additional element is then weighted using a closed-loop algorithm, to null out a single interfering signal.

10.4.2   Interference Nulling at the Receiver

If the presence of interference could be detected at the receiver and removed with a suitable filter, a genuine improvement in signal-to-interference ratio might be made.

Care must be taken that the filtering process itself does not degrade the intelligibility of speech. Individual nulls in the signal spectrum must not remove a significant fraction of the message power. For this technique to succeed, it is essential for the message signal to be relatively broadband, since it must be possible to identify the narrowband interference pattern unambiguously in the presence of the signal.

There are two basic techniques:

1   *Filter Bank Method*
The signal and interference are passed through a bank of identical filters, which cover the total signal bandwidth (see Figure 10.4). The mean signal power in each filter is measured and filters with outputs substantially above the mean

level are switched out. This technique requires the minimum of real time computation and adaptation and moreover could be carried out using analog techniques. It is, therefore, an attractive technique. However, careful design of the filters is needed since their group delay and amplitude characteristics must be accurately matched, particularly at the band edges. It is also desirable that each filter has a fairly narrow bandwidth (typically less than 100 Hz) in order to discriminate effectively against the interference. This implies a large number of complex filters (over 30 to cater for a 3 kHz bandwidth) and these are unlikely to be realisable by conventional analog techniques. The practicality of synthesising such a filter bank using digital techniques depends upon the processing power available.

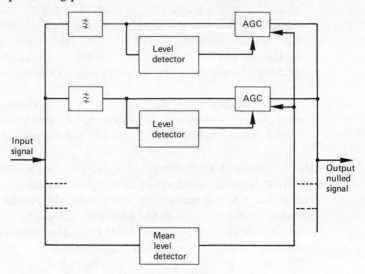

Fig. 10.4 *Filter bank interference nulling*

2 *Real Time Spectral Analysis and Selective Nulling*

The alternative approach to the filter bank is to place a small number of nulls in the message spectrum on the basis of a real time analysis of spectral occupancy. The individual filters used to generate the nulls may again need to be complex, but since only a small number is required, the complexity of the system need not be prohibitive.

The spectral analysis may be performed by a number of possible methods. In order of complexity they are

a) The use of a single frequency filter together with a swept frequency signal to determine the energy across the band.

b) The use of a delay line with a delay time proportional to frequency over the input bandwidth. If the input signal is mixed with a matched frequency sweep, the output from the delay line in the presence of narrowband interference will be a pulse for each interfering signal. The amplitude of each pulse is proportional to the interference amplitude and the output time is proportional to its frequency.

c) The use of a filter bank. Unlike the previous example of a filter bank the exact matching of the group delay characteristics is not necessary. Thus, the

filters for spectral analysis can be less complex than those used for interference nulling. As before, digital techniques can be used.

The optimum system to use for nulling the interference depends upon the
general implementation. If a largely digital approach is used, the most cost
effective technique is probably to store a set of fixed coefficients. These would
be used in a digital filter structure to implement notches at pre-set frequencies.

When largely analog techniques are used, a different approach is required,
since it is undesirable to manufacture a number of separate notch filters and
equalisers to cover the band. In this case, the best approach seems to be to have
the notch and equalisers at a fixed frequency and tuning is accomplished by
varying an external oscillator frequency (see Figure 10.5).

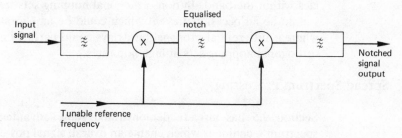

Fig. 10.5 *Interference notching system*

10.4.3 Frequency Agility

Modern high technology receivers and exciters with digitally controlled
synthesisers are capable of changing frequency several hundred times a second.
High-power transmitters and antenna system couplers are much slower, at
most a few times a second. However, state-of-the-art is progressing rapidly and
for lower-power wideband amplifiers with solid state couplers the tuning time
has been shortened dramatically.

The effect of an interfering signal on a given frequency can be reduced if the
dwell time of the wanted signal on that frequency is kept small. By moving
carrier frequency regularly to different parts of the spectrum, the wanted signal
will see only an average interference condition and will not be as susceptible to
any one strong interfering source. This is the principle adopted in frequency
hopping. A large number of predetermined channels are stored in a control
processor where they are progressively selected according to coding protocols.
The time spent in each channel (the dwell time) is as short as possible. In
practice best results are obtained if the band of frequencies employed is
relatively narrow (typically ±5% of the centre frequency) since the effects of
receiver and transmitter tuning and antenna coupling are then minimised.

The frequency used for transmitting the signal is changed regularly and
rapidly, normally many times a second. The actual frequencies used could be
pre-assigned or a band of frequencies centred on a known good frequency (or a
combination of the two). The choice of frequency at any one instant is normally
controlled pseudo-randomly. Control of the receiver local oscillator using the
same pattern as that used in the transmitter removes the wideband modulation
introduced.

Synchronisation of the system can be achieved in a number of ways. Since hopping rates at HF must be relatively low, synchronisation and accurate flywheeling can be achieved using readily available clock sources.

Interleaving of the data over many hops at the transmitter and the use of error detection and correction techniques can enable error-free transmission even under jammed conditions, but introduces inevitable delays over the link making it, possibly, unsuitable for speech transmission.

Frequency hopping techniques allow many users simultaneous use of the same frequency band by assigning each user his own code. Such systems can use non-orthogonal hopping sets where two users may use the same frequency at the same time each looking like an interfering signal to each other, or they can use orthogonal hopping sets where each user has a unique set of frequencies within the band. In non-orthogonal hopping sets frequency coincidence should be an occasional event which could be ameliorated by EDC coding, whereas differential propagation delays could partially negate the benefit of orthogonal hopping sets at high hop rates.

10.4.4   Spread Spectrum Processing

Section 9.5 has already demonstrated some advantages of using spread spectrum techniques which enable an overall signal power gain of $1/Dt_w$ to be achieved (the data elements are of duration $1/D$ and the chip width of the pseudo-random stream is t_w, as shown in Figure 10.6).

When the pseudo-noise signal is correlated with another signal, the result will be again a pseudo-noise signal, except for the unique case where the second signal is an exact replica of the first and is also in phase with it. Thus, in Figure 10.6, only if the sequence (c) is correlated with sequence (a) exactly in phase, will the original data stream (b) be recovered.

This property of the spread spectrum signal can be used to deal effectively with both broadband and narrowband interference. For broadband interference (Figure 10.7a), the correlation process picks out the broadband wanted signal, whilst keeping the interference broadband. For narrowband interference (Figure 10.7b) the correlation process at the receiver again provides a narrowband output for the wanted signal. However, at the same time it also spreads the interfering signal over the spread bandwidth by a similar process to the way in which the wanted signal was spread originally. The net effect of the spread spectrum process for either type of interference is therefore a much enhanced signal-to-interference ratio.

10.4.5   Interference Cancellation

Where conditions permit, the introduction of a locally generated electrical signal which acts as a 'cancelling' tone, inverse in phase to an audio interfering tone, can be used to reduce the effects of interference. To be effective, however, this must follow the interference automatically not only in frequency but also in phase and amplitude. (A failure by 180° in phasing of the cancelling tone in this method could result in enhancement of discrete interference by 6 dB!) However, the performance of noise cancelling microphones in simulated operations room environments is now extremely good.

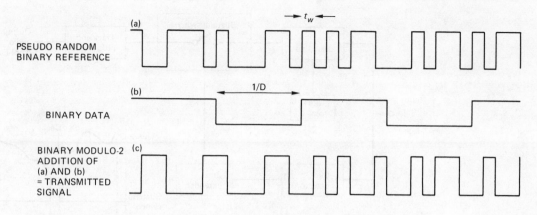

Fig. 10.6 *Data modulation of pseudo-random binary sequence*

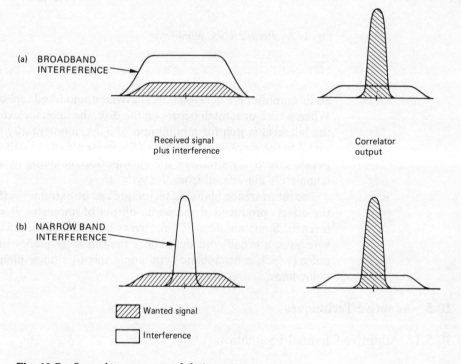

Fig. 10.7 *Spread spectrum modulation process*

10.4.6 Interference Blanking

Interference or noise blanking systems are now available which can operate on the received audio signal to remove the effects of impulsive electromagnetic interference. The principles can best be explained by recourse to the example of a record 'click' eliminator (see Figure 10.8). The disc signals are fed to the audio amplifier via a 3 ms delay line, a bilateral switch, and a track-and-hold circuit. Normally, the bilateral switch is closed, and the signal reaching the

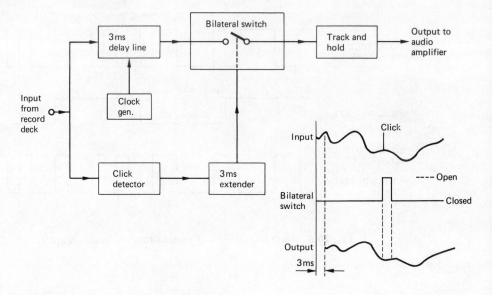

Fig. 10.8 *Record 'click' eliminator*

audio amplifier is a delayed but otherwise unmodified replica of the disc signal. When a click or scratch occurs on the disc, the detector/extender circuit opens the bilateral switch for a minimum of 3 ms, momentarily blanking the audio signal to the amplifier. Because of the presence of the delay line, the blanking period effectively straddles the click period, enabling its sound effects to be completely eliminated from the system.

The interference blanking technique can be extremely effective in removing the effects produced at the audio output of impulsive electromagnetic interference. Some amateur HF receivers have impulsive noise blanking circuits which use a broadband signal near the tuned frequency to identify the noise pulse (which is broadband) and apply this to a noise blanking circuit with a delay line.

10.5   Adaptive Techniques

10.5.1   Adaptive Channel Evaluation

The development of real time channel evaluation (RTCE) systems will enable the available capacity of the HF medium to be used more effectively in future by monitoring the shorter-term time fluctuations of a set of assigned HF channels. This RTCE data could be used in the control of an adaptive transmission rate system in which the transmission rate at any time could be matched more accurately to the available capacity. To implement this form of channel monitoring economically, it would be necessary to time/frequency multiplex RTCE probing signals and measurement periods with the traffic signals, possibly with the input data stream being buffered during the probing/measurement intervals.

One of the main disadvantages of RTCE systems as currently implemented is that they are separate from the communications system which they are designed to support; they require dedicated and expensive units which are comparable in cost to the units of the communications system itself.

In the context of the overall HF communications system, RTCE data could, in principle, be employed as a source of control information to assist in the adaptation of the following parameters[5]:

Transmitter power level
Frequency of operation
Bandwidth
Information rate
EDC algorithm
Modulation type and spectral format
Start time and duration of transmission
Antenna characteristics
Diversity combining algorithms.

To achieve the potential advantages, a number of major technical problems need to be solved in future HF designs.

It will be particularly important to develop RTCE techniques that can be integrated with the communications function of the system itself. Thus the radio sub-system and the RTCE sub-system should employ a common range of RF and processing equipment. System control algorithms must be capable of being fully automated.

Techniques will need to be developed to allow the transmission of reliable digitised speech over all types of HF paths, whilst adaptive or re-configurable signal generation and processing equipment is needed for use in variable capacity links. The design techniques applied should, of course, aim to minimise both system operation costs and maintenance costs.

10.5.2 Adaptive Detection Processes

Equalisers have already been discussed in section 9.4.3 in connection with sending fast serial data transmissions over HF links. Equalisers aim to cancel the effects of the channel before decoding the signal.

The algorithms used for setting tap coefficients range from simple gradient to steepest descent types, the latter including Kalman filtering[6], which is very difficult computationally and modified 'fast Kalman' algorithms which are easier to implement. An alternative technique to setting the tap coefficients is to send an isolated pulse between data messages and use this to estimate the sampled impulse response of the channel, and hence the tap coefficients. This approach[7] would need to be done 25 times a second to follow the varying channel, adding significant redundancy to the transmitted message.

An alternative approach to the adaptive equaliser is to include the effects of the channel in the decoding section. Obviously the detection process must be adaptive like the equaliser, and this can be done by including a channel estimator to model the HF channel. This is a detection process that operates on the received signal sequences, not just on each individual symbol. It offers a generally improved tolerance to Gaussian noise than could be obtained with a non-linear equaliser. The aim of the detection process is to determine the sequence of symbols which is most likely to have been transmitted from the received signal message. The Viterbi algorithm can perform maximum likelihood detection, but is not often used in practice because of its complexity. In order to make the algorithm usable in real time it is modified by reducing the number of possible sequences that it retains for each new symbol.

In addition to storing each sequence a figure of performance of that sequence (i.e. how likely), known as the cost, is also stored. When the whole sequence (or many symbols) have been received the vector representing the sequence with the least cost is used as the decoded symbol sequence. The cost associated with each vector is the sum of the errors squared in each possible symbol.

The above detection process would still require an adaptive equaliser before the detector to remove multipath. However, with further modification the detector can operate provided it knows the sampled impulse response of the channel. The latter could be found from sending isolated pulses down the channel, or more attractively by using a channel estimator. This operates upon the received symbols and the estimated detected symbols to provide the modified Viterbi detector with the sampled impulse response. This tracks the channel more simply than for the case with an adaptive equaliser[8].

10.5.3   Adaptive Receive Array Processing

The improvement in circuit reliability by using fixed or rotatable directional antennas at HF is well known. For ground-based systems an important area for the future is the use of antenna arrays which will become more common. The special case of interference nulling has already been addressed in section 10.4.1. VLSI techniques provide the processing power necessary to perform efficient wavefront analysis. Such a procedure can determine the constituent parameters of a multi-component wavefield from measurements of complex amplitudes of signals received in the elements of an antenna array. The dynamic nature of the HF medium necessitates the use of a fast processing capability to optimise the system performance. Most HF receive stations currently provide only limited receive directionality and rely on manual selection and operator judgement to choose a suitable antenna. The aim in future ground stations should be to provide automatic optimisation of receive antenna directivity by electronically steering the main beam of the antenna array so as to maximise signal-to-noise ratio.

An adaptive antenna array is designed to enhance the reception of desired signals in the presence of interference and deliberate jamming. Such systems have been built and tested in a variety of communications applications including land mobile, seaborne and airborne platforms. The utility of adaptive arrays for communication lies mainly in the ability to reject the undesired signals by virtue of their spatial properties.

Many adaptive arrays operate on the principle of maximising the signal-to-noise ratio or minimising the power in the undesired signal at the output of the array. This is achieved by subtracting a reference signal which matches the desired signal from the array output signal. Various techniques are available to perform these complex computations quickly and accurately. Some of the techniques can be implemented using digital, analog or a combination of analog and digital data processing methods.

An attractive possibility where the signal is digital is to proceed as follows: once the signal has been identified, select a number of additional elements around the array. Re-process the signal samples stored in the high speed temporary store using different combinations of weights so as to generate a number of beams. Select further array elements (discarding others) in an iterative process so as to obtain the best recovered signal, using the prior knowledge of the transmitted signal obtained in parallel from the signal identification process. A variant of the approach, if the position of the transmitter is known approximately, is to shorten the iteration by selecting elements and weights so as to generate a beam in the transmitter direction.

Estimation of the computational workload required for adaptive beamforming and nulling can only be approximate without detailed specification of the algorithms. The processing can be partitioned into two: antenna beamforming (in which linear combinations of in-phase and quadrature samples from each element must be generated at the full sampling rate) and adaptive weight calculation. The workload involved in beamforming is thus directly proportional to the number of array elements n ($3n$ instructions per sample is a reasonable estimate). Adaptation at worst could involve matrix inversions or equivalent operations, but needs only take place at a rate sufficient to cater for ionospheric fading, for example, at intervals of greater than 10 ms. However, the processing load associated with matrix inversion varies as the cube of the number elements (say $10n^3$ for a complex matrix). Even with a large safety factor to allow for computational inefficiencies, organisation and manipulation of data, the workload associated with adaptive beamforming is well within the capacity of modern digital signal processors.

10.5.4   Adaptive Modulation and Coding

Previous attempts at adaptive equalisation have proved to be too slow to follow changes in the propagation characteristics of the HF medium. Recent approaches, using coding as well as complex modulation schemes, rely on advanced technology for their successful implementation.

Codem[9] is a relatively recent, and somewhat unconventional approach. It uses a design of modulation and coding combined in such a way as to enhance the effectiveness of the overall communications system. The codem technique uses a form of channel measurement in the decoding process to enable an assessment to be made of the required coding complexity. The method of modulation is chosen to suit the code. The reliability of each bit is measured by the amplitude of the received tone, and an analog measure of the phase transition between successive bits. The reliability results are then used to make the decoding more effective. A significant performance improvement can be achieved without introducing an overall system delay beyond one frame. The effects of frequency selective fading can be reduced by the use of the error correcting code across the frequency band, since the error rate is essentially determined by the tones of lowest amplitude.

A different approach could use information received about the propagation medium in the data stream itself. The application of Kalman filtering techniques should enable the receiver to compensate the incoming data stream for the effects produced by the medium. Although it is becoming increasingly common when applied to navigation systems, the problem is much more extreme for an HF data communications system and considerable further work is necessary to produce a viable system.

The variable nature of the HF channel can also be exploited by use of embedded data encoding. In this case a single transmitted signal format contains several versions of the source data, each keyed at a different rate. The receiver decodes the signal at the highest data rate which the channel capacity would instantaneously allow. System control is accomplished by an ARQ arrangement. As an example, consider Figure 10.9 which shows three identical data streams each clocked at a different rate. At given intervals the receiver indicates to the transmitter which of the three data streams it has been able to receive successfully in a previous transmission interval. As a result of this feedback the transmitter re-aligns the data blocks over the subsequent transmission interval. This form of encoding has yet to be developed and implemented for HF applications[10].

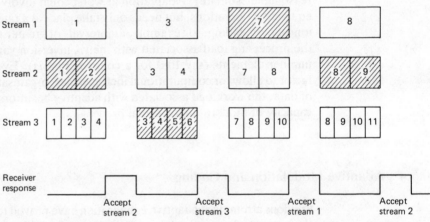

Fig. 10.9 *Embedded data encoding scheme*

Burst interference in the HF band can result in the loss of significant parts of a message. There will be increasing emphasis on the need to protect information whether it be from unauthorised users or from unwanted interference. This can be achieved by the use of advanced modulation techniques and encryption. The latter is most easily effected for voice signals if the speech waveforms are digitised, while the use of spread spectrum techniques to overcome unwanted interference is also readily implemented through the use of digital waveforms.

The advent of microelectronics has provided a means to develop systems that have built-in error detection and correction (EDC) which, until recently, would not have proved cost effective. Sophisticated EDC techniques employ powerful codes and interleaving to reduce error rates to acceptable levels; the effective data rate is thereby decreased. Powerful codes, such as Reed-Soloman, are now capable of correcting a high percentage of errors. VLSI techniques enable high levels of data interleaving to be used in order to distribute any bursts of errors into a random sequence when the data stream is de-interleaved at the receiver.

Interleaving of message bits over a period longer than the expected fading cycle is an effective way to distribute the bit errors, so that coding techniques can be used to correct them and to achieve reduction in error rates of several order of magnitude. This is, of course, at the expense of the information rate. For digital voice applications, there may be an unacceptable delay caused by error correction processing algorithms. A compromise must therefore be maintained between coding complexity, level of interleaving and tolerable processing delay.

10.6   Speech Processing

10.6.1   Voice Digitisation

Shannon's information theory predicts that the capacity of a 3 kHz channel with a good signal-to-noise ratio is around 30 kbit/s. In practice it is extremely difficult to approach this sort of data rate. For a 3 kHz HF channel operating over a sky wave link Chapter 9 has shown that problems of intersymbol interference limit serial data rate to two orders of magnitude less than the theoretical figure. However, adaptive equalisation techniques or parallel data modulation methods can substantially improve upon a few hundred bits/s. In the ground wave mode the passband of a 3 kHz HF SSB channel is approximately flat and subject mainly to degradation caused by noise and interference, which at times may be severe. This limits an achievable data rate in practice to a few thousand bits per second.

Since the 1960s pulse code modulation (PCM) at 64 kbit/s has been the worldwide standard for digitised speech. The resulting signal quality is almost indistinguishable from the original analog signal. PCM and other techniques such as delta modulation are relatively simple because they make little or no use of the fact that speech is redundant. Unfortunately the data rate constraints over an HF channel mean that the use of low cost and simple voice digitisation processes such as PCM are not feasible for HF applications.

Although the true information rate in speech is only about 50 bit/s no practical real time conversational speech digitisation devices operating at 50 bit/s exist. However, the development of compressed, real-time, digital voice coding techniques is progressing rapidly. New digital signal processing (DSP) and VLSI technology is allowing the compression techniques to be implemented cost effectively. Such techniques are becoming very useful in real time applications where bandwidth is scarce or valuable or where voice security with digital encryption is desired for communication over limited bandwidth facilities. These are just such conditions that apply to the HF band.

Techniques for transmitting digital voice fall into two broad categories:

a) Vocoders, producing data at typically 2400 bit/s.
b) Speech recognisers, producing data at much lower data rates but limited to specific word vocabularies.

10.6.2   Vocoders

Vocoders are devices which allow the low bit rate transmission and reception of speech. Vocoders work by attempting to describe not the signal itself directly, but those parameters of an idealised speech model which fully describe the fundamental speech information contained in the signal[11]. The broadband excitation signal from vocal sounds is modified by the complex-shaped cavity producing the sounds to give a spectrum with resonances. These resonances are known as *formants* and their frequency, amplitude and bandwidths determine the voiced sound produced. Digital codes, representing the features of the spoken words, are transmitted and the data is used to reconstruct speech at the receiving end of the link.

The earliest vocoders were channel vocoders, which use a filter bank in the analysis of the formant structure, and a series of controlled oscillators and filters to reconstruct it. Typically 20 filters might be used to give a good overall description of the four or five resonances in the spectrum of the formant structure.

More recent vocoders have used *linear predictive coding* and formant tracking techniques which have the potential of somewhat lower bandwidth. This method assumes that the vocal tract resonances can be modelled by a linear all-pole filter. A typical system, LPC10, uses ten poles, which equate to a maximum of five resonances. The pole locations, estimated by minimising the prediction error between the filter output and the speech signal, are transformed into digits that are used to define the spectrum. LPC10 uses frames of 22.5 ms duration, encoding each frame into 54 bits, hence producing 2400 bit/s. On reception the data sequence is segmented into frames and the value of each parameter determined.

Vocoders are able to provide a 'reasonable' speech quality with large vocabularies. However, since the speech is synthesised using an idealised model of the speech production process, it is not a direct copy of the original. It is not surprising therefore that on first listening to vocoded speech it sounds strange and rather difficult to understand. A few hours familiarity with a vocoder improves the perception radically. The intelligibility of vocoded speech by a 'trained' speaker/listener pair is generally extremely good. It is possible to alter the characteristics of the replayed speech if desired. This feature can be used to advantage in order to enhance intelligibility in a noisy environment by emphasising certain aspects of the speech waveform.

10.6.3   Speech Recognisers

Spoken word recognition devices are now available which can represent a word by a few binary digits. This offers the possibility of using much narrower band channels which could give a considerable reduction in the noise level in HF voice links. In a digital word encoding system the number of bits needed to represent each word depends entirely on the size of the vocabulary and for many operational uses this can be quite small. For example, a 256-word vocabulary would require an 8-bit address for each word. A speech rate of 2 words per second gives 16 bits per second which would be at least doubled (say 40 bits per sec) if EDC, synchronisation and framing bits were to be added.

Transmission could then be effected over a channel with a very narrow bandwidth.

All speech recognition systems can be considered as belonging to one of three classes:

Isolated Word Recognition
Connected Word Recognition
Continuous Speech Recognition.

The latter two classes are known generically as *connected speech*. However, it is often difficult to identify a particular approach as belonging to any one class. For example, an isolated word recogniser may be capable of recognising short phases and so might be described as a connected word recognition system. Similarly, a continuous speech recognition system is equally capable of operating on connected or isolated words.

1   *Isolated Word Recognition*

Speech recognition systems which require a short pause of at least 100 ms between each utterance are classed as isolated word recognisers. However, 'word' in this context should be taken to mean anything from single words to short phrases of one or more seconds duration. Nearly all of these systems have to be trained to recognise the operator's voice characteristics before use.

Isolated word recognisers with vocabularies of hundreds of words are now becoming available in commercial form. Most of these are speaker-dependent, although a few suppliers are beginning to offer speaker-independent systems with very limited (a few tens of words) vocabularies. High accuracies are claimed for small word vocabularies under relatively quiet conditions.

2   *Connected Speech Recognition*

Connected word recognisers are often considered to belong to the same class as continuous speech recognition systems although there are significant differences between them. *Connected word recognition* systems may be referred to as those which recognise each word in a naturally spoken phrase bounded by pauses of greater than 100 ms duration.

Currently available connected word recognition systems are limited to relatively short phrases and require training to recognise the operator's voice characteristics. Recent developments in speech recognition allow an operator to communicate with machines using natural, conversational speech. This approach is known as *continuous speech recognition* and usually operates on a word spotting basis. Thus, continuous speech recognition systems do not attempt to recognise every utterance but pick out words from continuous speech which are in the chosen vocabulary.

During normal speech it is difficult for a machine to determine where one word ends and another begins. It is therefore possible for the recognition system to form spurious words made up from the end of one word and the start of the next, or to spot words within other words. In addition, the acoustic characteristics of sounds and words exhibit much greater variability in continuous speech, depending on the context, compared with words spoken in isolation.

Systems that are currently available tend to require each user to input their own voice 'templates' via cassette or read only memory. They are thus speaker-dependent, with extendable word storage to many hundreds of words. Programmable word sequence rules may also be incorporated. Connected word recognition of utterances of up to 8 seconds are becoming common. Many systems are still physically large but generally use readily available components. There seems no reason why more compact and rugged versions, for example suitable for airborne use, should not be produced. Speech recognition chip sets are also becoming available to perform linear predictive coding feature extraction on a digitised audio signal.

10.6.4 Benefits of Speech Processing

The speech processing opportunities provided by current VLSI technology offer a number of significant advantages to the user of HF communications.

1 *Bandwidth Reduction*
One of the major benefits of the use of speech processing on a communications channel is the reduced bandwidth that is required. A vocoder system with a data rate of 2000 bit/s is current technology; 200 bit/s is feasible but the resulting intelligibility is open to question. Any method of using narrower bandwidths whilst retaining satisfactory speech quality should offer at least some alleviation of the noise problem. However, the use of narrowband vocoders makes speech extremely susceptible to corruption caused by data errors from in-band interference. Nevertheless, the reduced bandwidth gives increased immunity from jamming and broadband interference. Alternatively the spare capacity of a higher bandwidth can be harnessed for additional coding or security.

2 *Message Compression*
The reduced data rate of digitised speech, and in particular source-coded speech, can have other benefits as well as reducing the bandwidth. One alternative is to store periods of speech and transmit them as bursts of data. About one minute's worth of source coded speech could be transmitted in about one second over a normal communications channel. This message compression can make the communications link much more secure, and also reduces spectrum pollution.

3 *Audio Noise Reduction*
Speech recognition and encoding can alleviate the effects of background audio noise in two respects. In the actual recognition process a hard decision is made on the word being spoken. High interfering acoustic noise at the microphone may make this decision difficult, and so result occasionally in a word recognition error. However, the code produced will contain no information on the noise which will therefore have been effectively completely filtered out. At the receiving end of the link (after EDC decoding has been carried out) the resulting code has then to be converted back into speech, using a speech synthesiser; once again a hard decision has to be made on the coded word. As far as the receiving operator is concerned this system provides a noise-free acoustic output.

4 Security

There are two broad classes of secure speech systems which have been used, namely scramblers and those using vocoders. *Scramblers*, which attempt to directly separate the signal into small elements and mix them in an apparently random way, cannot be considered as totally secure in the military sense, although for most commercial purposes they provide a satisfactory degree of privacy.

However, the transformation by use of a *vocoder* of an analog speech signal into a digital data stream, which is passed through a digital encryption device, would produce a resultant data stream which is, to all intents and purposes, completely random. The transmission of such a data stream, using an appropriate modem, is therefore secure. In information theory terms it has no redundancy, each bit is completely independent and unrelated to all other bits. Such a system is shown schematically in Figure 10.10.

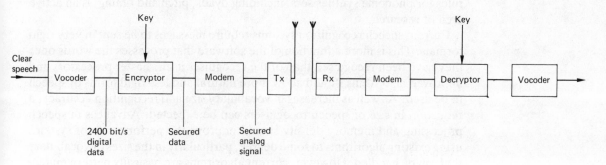

Fig. 10.10 *Vocoder used in a secure speech system*

In speech recognition systems code numbers instead of words are transmitted, and the relationship between the codes and words can be assigned arbitrarily. At the expense of some increase in bandwidth, additional encryption of a wide variety of forms, whether fixed or time dependent, can be applied. The overall result can be a highly encrypted signal which still only requires a comparatively low bandwidth.

5 Alternative Outputs

Word encoding systems have the further advantage that decoding does not necessarily have to result in a speech output. In many circumstances the ability to read a message on a visual display unit where it can be retained for a short time, rather than having to listen to it, can have obvious advantages.

6 Radio Control

Speech recognition can be used as means of controlling radio communications, a feature which is particularly useful where there is a large number of controls in a rather cramped space, as in a modern aircraft cockpit. In current systems controls for radios are often inconveniently located which may make them difficult to operate. It may also mean that they are not easy to see without excessive head movement by the pilot. Vibration and *g* forces within the aircraft can also make adjustment of controls rather difficult.

Control of the radios by direct voice input overcomes these problems, in that the pilot does not need to be able to reach them easily nor does he need to look down to re-tune a radio. A speech recogniser combined with a comparatively simple processor could repond to station names instead of frequencies or channel numbers.

10.6.5   Future Trends

Future work on vocoders will concentrate on reduced bandwidth and increased intelligibility. The use of vocoders based on phoneme recognition and synthesis is attractive for bit rates down to 200 bit/s but gives less natural-sounding speech. The problem is that the articulation of a phoneme depends on the phonemes before and after it, so mere concatenation of phonemes is not satisfactory. One solution to this co-articulation problem is to introduce transitional sound (called dyads) from one phoneme to the next. The generation of rules for phoneme synthesisers, including dyads, pitch and timing, is an active area of research.

Current speech recognition systems require messages to be sent in very rigid formats. This is more a function of the software that processes the words once they have been recognised than of the recogniser itself. Faster processors and improved algorithms should allow more natural and less rigid forms of speech to be used. As well as increases in vocabulary size and recognition accuracy, a reduction in size of speech recognisers can be expected. Advances in speech processing and memory density should improve the performance of systems using existing algorithms to some degree, particularly in the sizes of vocabulary that can be handled. However, current algorithms are basically pattern matching, whereas humans interpret speech as much by the context as by the speech pattern. Thus current machines are incapable of distinguishing between the phrases 'Recognise speech' and 'Wreck a nice beach'. A human, however, can easily tell the difference from the context.

In the future, intelligent knowledge-based systems (IKBS) may be able to give speech recognisers the ability to consider the context of a phrase. The context may cover a variety of possible subjects such as previous phrases, or the physical situation. For example, an airborne system might be able to infer the meaning of a phrase from the current aircraft activity. These areas, however, are dependent on the availability both of suitable IKBS machines and of the necessary algorithms. There is no doubt that they will make their mark on the capabilities of future HF communications systems.

References

Chapter 2

[1] International Standards Organisation, *Information Processing Systems, Open Systems Interconnection, Basic Reference Model*, International Standard ISO/DIS 7498

Chapter 3

[1] CCIR, Conclusions of the interim meeting of Study Group 5, *Propagation in non-ionised media*, ITU, Geneva, 1980

[2] K Bullington, *Proc. I.R.E.* **35**, pp 1122-1136, 1947

[3] A N Sommerfeld, The propagation of waves in wireless telegraphy, *Ann. Phys, Series 4*, **28**, pp 665, 1909

[4] J R Wait, *Electromagnetic Waves in Stratified Media*, Pergamon Press, 1970

[5] CCIR, *Ground wave propagation in an exponential atmosphere*, Report 714, Study Group 5, ITU, Kyoto, 1978

[6] P David and J Voge, *Propagation of Waves*, Pergamon Press

[7] CCIR, *Propagation by diffraction*, Report 715, Study Group 5, ITU, Kyoto, 1978

[8] T Tamir, On radio propagation in forest environments, *IEEE Trans. Ant. Prop.* **AP-15**, pp 806-817, 1967

[9] P Beckmann and A Spizzichino, *The Scattering of Electromagnetic Waves from Rough Surfaces*, Pergamon Press, London, 1963

[10] H R Reed and C M Russell, *Ultra High Frequency Propagation*, Wiley, New York, 1953

[11] P Beckmann, Shadowing of random rough surfaces, *IEEE Trans. Ant. Prop.* **AP-13**, pp 384-388, 1965

[12] K Furutsu, Wave propagation over an irregular terrain, *J.Radio. Res. Labs*, Japan, **6**, pp 23, 1959

[13] R S Kirby, Obstacle gain measurements over Pikes Peak at 60 to 1046 MHz, *Proc IRE* **43**, pp 1467-1472, 1955

[14] CCIR, *Influence of terrain irregularities and vegetation on tropospheric propagation*, Report 236-4, ITU, Kyoto, 1978

[15] T Tamir, Radio wave propagation along mixed paths in forest environments, *IEEE Trans. Ant. Prop.* **AP-25**, pp 471-477, 1977

[16] R H Ott, An alternative integral equation for propagation over irregular terrain, *Radio Science* **6**, pp 429-435, 1971

[17] CCIR, *Atlas of Ground Wave Propagation Curves for Frequencies between 30 MHz and 10000 MHz*, ITU, Geneva, 1959

[18] H Bremmer, *Terrestrial Radio Waves*, Elsevier, 1949

[19] B R Bean and E J Dutton, *Radio Meteorology*, Dover Publications, New York, 1968

[20] CCIR, *Reference atmosphere for refraction*, Recommendation 369-2, ITU, Kyoto, 1978

[21] S Rotheram, Ground wave propagation, *Proc IEE*, Part F, pp 275-295, Oct 1981

[22] CCIR, *Ground wave propagation curves for frequencies between 10 kHz and 30 MHz*, Recommendation 368-3, Study Group 5, ITU, Kyoto, 1978

[23] CCIR, *Electrical characteristics of the surface of the Earth*, Recommendation 527, Study Group 5, ITU, Kyoto, 1978

[24] G Millington, Ground wave propagation over an inhomogeneous smooth earth, *Proc IEE*, Part III,**96**, pp 53, 1949

[25] D E Barrick, Theory of HF and VHF propagation across rough sea, *Radio Science* **6**, pp 517-533, 1971

[26] A Picquenard, *Radio Wave Propagation*, Macmillan, London, 1974

[27] O M Murray, Attenuation due to trees in the VHF/UHF bands, *The Marconi Review* **37**, pp 41-50, 1974

Chapter 4

[1] CCIR, *Ionospheric properties*, Report 356, ITU, Kyoto, 1978

[2] K Davies, *Ionospheric Radio Propagation*, Dover Publications, New York, 1966

[3] K G Budden, *The Propagation of Radio Waves*, Cambridge University Press, 1985

[4] N M Maslin, High data rate transmissions over HF links, *Radio and Electronic Engineer* **52**, pp 75-87, February 1982

[5] D R Sloggett, Improving the reliability of HF data transmissions, *IEE Colloquium Digest* **48** (Recent advances in HF communications systems and techniques), 1979

[6] K Davies and D M Baker, On frequency variation of ionospherically propagated HF radio signals, *Radio Science* **1**, pp 545-556, 1966

[7] C C Watterson, Experimental confirmation of an HF channel model, *IEEE Trans. Com. Tech.* **COM-18**, pp 792-803, 1970

[8] J M Headrick and M I Skolnik, Over-the-horizon radar in the HF band, *Proc. IEEE* **62**, pp 664-674, 1974

[9] R A Shepherd and J B Lomax, Frequency spread in ionospheric radio propagation, *IEEE Trans. Com. Tech.* **COM-15**, pp 268-275, 1967

[10] A Malaga and R E McIntosh, Delay and Doppler power spectra in a fading ionospheric reflection channel, *Radio Science* **14**, pp 859-873, 1978

[11] H H Inston, Dispersion of HF pulses by ionospheric reflection, *Proc. IEE* **116**, pp 1789-1793, 1969

[12] J W Herbstreit and W Q Crichlow, Measurement of the attenuation of radio signals by jungles, *Radio Science* **68D**, pp 930, 1964

[13] G H Hagn and W R Vincent, Comments on the performance of selected low-power HF radio sets in the tropics, *IEEE Trans. Veh. Tech.* **VT-23**, pp 55-58, 1974

[14] CCIR, *Second CCIR computer-based interim method for estimating sky wave field strength and transmission loss at frequencies between 2 and 30 MHz*, Supplement to Report 252, Doc 6/1070-E, ITU, 1978

Chapter 5

[1] ITT, *Reference Data for Radio Engineers*, Howard Sams, Indianapolis, 1977

[2] K Davies, *Ionospheric Radio Propagation*, Dover Publications, New York, 1966

[3] CCIR, *World distribution and characteristics of atmospheric radio noise*, Report 322-1, ITU, Geneva, 1974

[4] CCIR, *Man-made radio noise*, Report 258-3, ITU, Geneva, 1978

[5] H V Cottony and J R Johler, Cosmic radio noise intensities in the VHF band, *Proc IRE* **40**, pp 1053-1060, 1952

[6] CCIR, *Second CCIR computer-based interim method for estimating sky wave field strength and transmission loss at frequencies between 2 and 30 MHz*, Supplement to Report 252, Doc 6/1070-E, ITU, 1978

[7] S Dutta and G F Gott, HF spectral occupancy, IEE Conference Publication 206 on *HF Communication Systems and Techniques*, 1982

[8] R A Cottrell, An automatic HF channel monitoring system, *IEE Colloquium Digest* **48** (Recent advances in HF communications systems and techniques), 1979

[9] G F Gott, N F Wong and S Dutta, Occupancy measurements across the entire HF spectrum, *AGARD Conference Proceedings CP-332*, 1982

[10] L M Posa et al, Azimuthal variation of measured HF noise, *IEEE Trans. EMC* **EMC-14**, Feb 1972

[11] G F Gott and M J D Staniforth, Characteristics of interfering signals in aeronautical HF voice channels, *Proc IEE* **125**, pp 1208-1212, 1978

[12] M Darnell, An HF data modem with in-band frequency agility, *IEE Colloquium Digest* **48** (Recent advances in HF communications systems and techniques), 1979

[13] N J Carter and J M Thomson, Susceptibility testing of airborne equipment – the way ahead, *2nd symposium and Technical exhibition on EMC*, UK, 1977

[14] British Standards Institution, *General Requirements for Equipment for Use in Aircraft* (BS 3G 100) Part 4, Section 2, Electromagnetic interference at radio and audio frequencies, London, 1973

[15] N M Maslin, HF communications to small low-flying aircraft, *AGARD CP-263: Special topics in HF propagation*, NATO, 1979

Chapter 6

[1] R D Hunsucker and H F Bates, Survey of polar and auroral region effects on HF propagation, *Radio Science* **4**, pp 347-365, 1969

[2] D E Watt-Carter, Survey of aerials and distribution techniques in the HF fixed service, *Proc IEE* **110**, July/Sept 1963

[3] D C Bunday, Noise equalisation in HF receiving systems, *Radio and Electronic Engineer* **47**, May 1977

[4] CCIR, *Diversity reception*, Report 327-3, ITU, Geneva, 1982

[5] G L Grisdale et al., Fading of long-distance radio messages and comparison of space and polarisation diversity reception in the 6-18 MHz range, *Proc. IEE*, pp 39-51, Jan 1957

[6] CCIR, *Bandwidths, signal-to-noise ratio and fading allowances in complete systems*, Recommendation 339-3, ITU, Geneva, 1974

[7] N M Maslin, Assessing the circuit reliability of an HF sky wave air-ground link, *Radio and Electronic Engineer* **48**, pp 493-503, 1978

[8] CCIR, *Man-made radio noise*, Report 258-3, ITU, Geneva, 1978

[9] H Jasik, *Antenna Engineering Handbook*, McGraw Hill, 1961

[10] M N Sweeting and Q V Davis, Electrically short HF antenna systems, *AGARD CP-263: Special topics in HF propagation*, NATO, 1979

[11] CCIR, *Second CCIR computer-based interim method for estimating sky wave field strength and transmission loss at frequencies between 2 and 30 MHz*, Supplement to Report 252-2, Doc 6/1070-E, ITU, 1978

[12] N M Maslin, Assessment of HF communications reliability, *AGARD CP-263: Special topics in HF propagation*, NATO, 1979

Chapter 7

[1] N A D Pavey, Radiation characteristics of HF notch aerials installed in small aircraft, *AGARD CP-139*, NATO, 1973

[2] B Burgess, The role of HF in air-ground communications – an overview, *AGARD CP-263: Special topics in HF propagation*, NATO, 1979

[3] N M Maslin, Assessing the circuit reliability of an HF sky wave air-ground link, *Radio and Electronic Engineer* **48**, 493-503, 1978

[4] N M Maslin, Assessment of HF communications reliability, *AGARD CP-263: Special topics in HF propagation*, NATO, 1979

[5] N M Maslin, HF communications to small low-flying aircraft, *AGARD CP-263: Special topics in HF propagation*, NATO, 1979

[6] F Lied, *High frequency radio communications, with emphasis on polar problems*, AGARDOGRAPH No 104, Technivision, Maidenhead, 1967

Chapter 8

[1] K Davies, *Ionospheric Radio Propagation*, Dover Publications, New York, 1966

[2] C M Rush et al., The relative daily variability of f_oF2 and h_mF2 and their implications for HF radio propagation, *Radio Science* **9**, pp 749-756, 1974

[3] J W King and A J Slater, Errors in predicted values of f_oF2 and h_mF2 compared with the day-to-day variability, *Telecommunication Journal* **40**, pp 766-770, 1973

[4] CCIR, *Second CCIR computer-based interim method for estimating sky wave field strength and transmission loss at frequencies between 2 and 30 MHz*, Supplement to Report 252-2, Doc 6/1070E, ITU, March 1978

[5] J D Milsom, HF sky wave links and frequency management in Europe, *STC Conference Proceedings: Towards improved HF communications in the European Environment*, May 1983

[6] CCIR, *Real time channel evaluation of ionospheric radio circuits*, Provisional report AK/6, October 1981

[7] M Darnell, Real time channel evaluation, *AGARD Lecture Series LS-127: Modern HF communications*, NATO, 1983

[8] M Darnell, HF system design principles, *AGARD Lecture Series LS-127: Modern HF communications*, NATO, 1983

[9] D C Coll and J R Storey, Ionospheric sounding using coded pulse signals, *Radio Science* **69D**, pp 1155-1159, 1964

[10] G H Barry and R B Fenwick, Extra-terrestrial and ionospheric sounding with synthesised frequency sweeps, *Hewlett-Packard J.* **16** no. 11, pp 8-12, 1965

Chapter 9

[1] C E Shannon, A mathematical theory of communication, *Bell Syst. Tech. J.* **27**, pp 379-424, pp 623-657, 1948

[2] H B Law, The detectability of fading radiotelegraph signals in noise, *Proc IEE* **104B**, pp 130, 1957

[3] J D Ralphs, *Principles and Practice of Multi-frequency Telegraphy*, IEE Telecommunications Series No 11, Peter Peregrinus Ltd, London, 1985

[4] F Amoroso, Pulse and spectrum manipulation in the MSK format, *IEEE Trans. Com.* **COM-24**, pp 381, 1976

[5] CCIR, *Prediction of the performance of telegraph systems in terms of bandwidth and signal-to-noise ratio in complete systems*, Report 195, ITU, Geneva, 1982

[6] CCIR, *Voice-frequency telegraphy on radio circuits*, Recommendation 106-1, ITU, Geneva, 1982

[7] CCIR, *Performance of telegraph systems on HF radio circuits*, Report 345-2, ITU, Geneva, 1982

[8] CCIR, *Diversity reception*, Report 327-3, ITU, Geneva, 1982

[9] CCIR, *Usable sensitivity of radio telegraphy receivers in the presence of quasi-impulse interference*, Report 183-3, ITU, Geneva, 1982

[10] H Nyquist, Certain topics in telegraph transmission theory, *Trans AIEE* **47**, pp 617-644, April 1928

[11] M L Doelz et al., Binary data transmission techniques for linear systems, *Proc IRE* **45**, pp 656-661, 1957

[12] G C Porter, Error distribution and diversity performance of a frequency differential PSK HF modem, *IEEE Trans. Com. Tech.* **COM-16**, pp 567-575, 1968

[13] B Goldberg, 300 kHz–30 MHz MF/HF, *IEEE Trans. Com. Tech.* **COM-14**, pp 767-784, 1966

[14] M S Zimmerman and A L Kirsch, The AN/GSC-10 (KATHRYN) variable-rate data modem for HF radio, *IEEE Trans. Com. Tech.* **COM-15**, pp 197-204, 1967

[15] M J Di Toro, Communication in time-frequency spread media using adaptive equalisation, *Proc IEEE* **56**, pp 1653-79, 1968

[16] N M Maslin, High data rate transmissions over HF links, *Radio and Electronic Engineer* **52** pp 75-87, February 1982

[17] G F Gott and J P Newsome, HF data transmission using chirp signals, *Proc IEE* **118**, pp 1162-1166, 1971

[18] R A Scholtz, The spread spectrum concept, *IEEE Trans. Com.* **COM-25**, pp 748-755, 1977

[19] A J Viterbi, Convolutional codes and their performance in communication systems, *IEEE Trans. Com.* **COM-19**, pp 751-772, 1971

[20] R J McEliece, *The Theory of Information and Coding*, Addison-Wesley, 1977

[21] R M F Goodman et al., Soft decision error-correction coding schemes for HF transmission, *IEE Colloquium Digest* **48** (Recent advances in HF communications systems and techniques), 1979

[22] B Maranda and C Leung, Block error performance of non-coherent FSK modulation in Rayleigh fading channels, *IEEE Trans. Com.* **COM-32**, pp 206-209, 1984

[23] G F Gott and B Hillam, The improvement of slow rate FSK by frequency agility and coding, *IEE Colloquium Digest* **48** (Recent advances in HF communications systems and techniques), 1979

[24] M Schwartz, W R Bennett and S Stein, *Communication Systems and Techniques*, McGraw-Hill, Inter-University Electronics Series, Volume 4, 1966.

Chapter 10

[1] J Dyer, High frequency receiver design II, *Radio and Electronics World*, pp 30-35, July 1983

[2] R W J Awcock, The lincompex system, *Point to Point Communications*, pp 130-142, July 1968

[3] CCIR, Recommendation 455-1, ITU Geneva, 1974

[4] J A Betts, *Signal Processing, Modulation and Noise*, Hodder and Stoughton, London, 1970

[5] M Darnell, Real time channel evaluation, *AGARD LS-127: Modern HF Communications*, NATO, 1983

[6] D D Falconer and L Ljung, Application of fast Kalman estimation to adaptive equalisation, *IEEE Trans. Com.* **COM-26**, pp 1439-1446, 1978

[7] W Hodgkiss and L F Turner, Practical equalisation and synchronisation strategies for use in serial data transmission over HF links, *Radio and Electronic Engineer* **53**, pp 141-146, 1983.

[8] A P Clark and H Y Najdi, Detection process of a 9600 bps serial modem for HF radio links, *Proc IEE* **130**, pp 368-376, 1983

[9] D Chase, A combined coding and modulation approach for communication over dispersive channels, *IEEE Trans Com.* **COM-21**, pp 159-175, 1973.

[10] M Darnell, HF system design principles, *AGARD LS-127: Modern HF Communications*, NATO, 1983.

[11] P A T Hellen, Military secure speech using HF, *Electronics and Power*, pp 232-238, IEE, March 1985.

Index